THE NEW MIDDLE AGES

BONNIE WHEELER, *Series Editor*

The New Middle Ages is a series dedicated to pluridisciplinary studies of medieval cultures, with particular emphasis on recuperating women's history and on feminist and gender analyses. This peer-reviewed series includes both scholarly monographs and essay collections.

PUBLISHED BY PALGRAVE:

Women in the Medieval Islamic World: Power, Patronage, and Piety
  edited by Gavin R. G. Hambly

The Ethics of Nature in the Middle Ages: On Boccaccio's Poetaphysics
  by Gregory B. Stone

Presence and Presentation: Women in the Chinese Literati Tradition
  edited by Sherry J. Mou

The Lost Love Letters of Heloise and Abelard: Perceptions of Dialogue in Twelfth-Century France
  by Constant J. Mews

Understanding Scholastic Thought with Foucault
  by Philipp W. Rosemann

For Her Good Estate: The Life of Elizabeth de Burgh
  by Frances A. Underhill

Constructions of Widowhood and Virginity in the Middle Ages
  edited by Cindy L. Carlson and Angela Jane Weisl

Motherhood and Mothering in Anglo-Saxon England
  by Mary Dockray-Miller

Listening to Heloise: The Voice of a Twelfth-Century Woman
  edited by Bonnie Wheeler

The Postcolonial Middle Ages
  edited by Jeffrey Jerome Cohen

Chaucer's Pardoner and Gender Theory: Bodies of Discourse
  by Robert S. Sturges

Crossing the Bridge: Comparative Essays on Medieval European and Heian Japanese Women Writers
  edited by Barbara Stevenson and Cynthia Ho

Engaging Words: The Culture of Reading in the Later Middle Ages
  by Laurel Amtower

Robes and Honor: The Medieval World of Investiture
  edited by Stewart Gordon

Representing Rape in Medieval and Early Modern Literature
  edited by Elizabeth Robertson and Christine M. Rose

Same Sex Love and Desire among Women in the Middle Ages
  edited by Francesca Canadé Sautman and Pamela Sheingorn

Sight and Embodiment in the Middle Ages: Ocular Desires
  by Suzannah Biernoff

Listen, Daughter: The Speculum Virginum and the Formation of Religious Women in the Middle Ages
  edited by Constant J. Mews

Science, the Singular, and the Question of Theology
  by Richard A. Lee, Jr.

Gender in Debate from the Early Middle Ages to the Renaissance
  edited by Thelma S. Fenster and Clare A. Lees

C153838690

WITHDRAWN

ENCOUNTERING MEDIEVAL TEXTILES AND DRESS

OBJECTS, TEXTS, IMAGES

Edited by

Désirée G. Koslin
and
Janet E. Snyder

To Our Parents

ENCOUNTERING MEDIEVAL TEXTILES AND DRESS
Copyright © Désirée G. Koslin and Janet E. Snyder, 2002.
All rights reserved.

First published in hardcover in 2002 by PALGRAVE MACMILLAN® in
the US—a division of St. Martin's Press LLC, 175 Fifth Avenue, New
York, NY 10010.

Where this book is distributed in the UK, Europe and the rest of the
world, this is by Palgrave Macmillan, a division of Macmillan Publishers
Limited, registered in England, company number 785998, of Houndmills,
Basingstoke, Hampshire RG21 6XS.

Palgrave Macmillan is the global academic imprint of the above
companies and has companies and representatives throughout the world.

Palgrave® and Macmillan® are registered trademarks in the United States,
the United Kingdom, Europe and other countries.

ISBN-13: 978-0-230-60235-9
ISBN-10: 0-230-60235-5

Library of Congress Cataloging-in-Publication Data
Encountering medieval textiles and dress : objects, texts, images /
edited by Désirée G. Koslin and Janet E. Snyder.
    p.   cm—(The new Middle Ages)
    Includes bibliographical references and index.
    ISBN 0–230–60235–5
    1. Costume—History—Medieval, 500–1500.   2. Textile fabrics,
Medieval.   3. Costume in art.   4. Costume—Symbolic aspects.
I. Koslin, Désirée G., 1944–   II. Snyder, Janet E.   III. New Middle
Ages (Palgrave (Firm))

GT575.E53   2002
391'.009'02-dc21

                                                        2002068444

A catalogue record of the book is available from the British Library.

Design by Letra Libre, Inc.

First PALGRAVE MACMILLAN paperback edition: December 2008
10 9 8 7 6 5 4 3 2 1
Printed in the United States of America.

Transferred to Digital Printing 2008

KENT LIBRARIES
AND ARCHIVES

C 153838690

Askews

CONTENTS

SERIES EDITOR'S FOREWORD

The *New Middle Ages* contributes to lively transdisciplinary conversations in medieval cultural studies through its scholarly monographs and essay collections. This series provides new work in a contemporary idiom about precise (if often diverse) practices, expressions, and ideologies in the Middle Ages. This volume, the thirty-first in the series, continues a concern expressed in an earlier series volume, *Robes and Honor*, edited by Stewart Gordon, with ways that material culture, in this case ceremonial dress, encodes but also ambiguates significant cultural symbolics. That volume invited us to think about robing as a "ceremonial metalanguage," but they also remind us that robes themselves as well as robing ceremonials have particular, local resonance. In *Encountering Medieval Textiles and Dress: Objects, Texts, Images,* editors Désirée Koslin and Janet Synder have assembled a rich range of surviving examples of dressing—across time, in different media, up and down the social ladder, across professions, and between genders—and the essays in this volume delineate the details while also interrogating the relation of the represented to the "real." We see keenly how in clothing alone the users/representers embody cultures, with a touch of the antique here, a foray into the foreign exotic there, and a constant consciousness of the body as the basic site for human display.

Bonnie Wheeler
Southern Methodist University

ACKNOWLEDGMENTS

Our interest in the material culture of the Middle Ages and its ramifications for human appearance evolved from our separate and parallel lifetime immersions in the study of dress and textiles and the meanings conveyed through their use. We are indebted to Jonathan J. G. Alexander, Institute of Fine Arts, New York University, and to Stephen Murray, Columbia University, who supported our respective dissertation research topics. We also warmly wish to thank the curators and museum professionals at the Metropolitan Museum of Art and the Pierpont Morgan Library who continue to encourage our investigations.

We are grateful to the many friends, colleagues, and acquaintances who listened and responded to questions and challenges with interest and patience. Our appreciation also goes to those who joined the discussions each year since 1995 in our sessions at The International Medieval Congress at Western Michigan University, Kalamazoo, Michigan, where many of the papers that appear in this volume received their first presentations.

Special thanks are due to Bonnie Wheeler, series editor of the New Middle Ages series at Palgrave, to William T. Clark, Queens' College, City University of New York, and to Danielle Johnson, the Limestone Sculpture Provenance Project, Paris, whose mentorship led to this project. We thank Dr. Laura Hodges for her generous and attentive reading of the manuscript. We owe gratitude to Kristi Long and Meg Weaver of Palgrave whose cheerful and professional expertise sustained us from the earliest to the final phases of publication.

We bask in the warmth of special friendships, sorely tried while growing stronger, that have carried us through to the finish. Lastly, but decidedly foremost, we applaud our talented authors who worked with dispatch and diligence on their contributions, making this book everything we wished for at the outset.

ILLUSTRATIONS

INTRODUCTION

Désirée Koslin and Janet Snyder

Audiences interested in the attire, dress, and textiles of the Middle Ages are aware of the importance of, as well as the challenges, in this field of study. It is widely recognized that medieval society depended on clothing codes and prestigious textile furnishings for signs of identity as well as the actual economic underpinnings of society. The evidence for these phenomena, however, is scant and embedded in the greater context of the surviving material from the period. Furthermore, between these sources and us lie several hundred years of interventions that have added facts and fiction, interpretation and alteration in an ongoing, multilayered process of change involving ideas about the culture of what we call the "Middle Ages."

Today the study of dress and textiles, undertaken through surviving objects and through representations in art, literature, and cultural commodities, is recognized as significant and is no longer "marginalized" in the academy. Indeed, scholars in a wide range of disciplines have taken it up as an area of specialization in their discrete fields, and at least two scholarly journals publish writings on the continued search for identity of the discipline.[1] Most of us also owe a debt for our methodological tools to the seminal contributions of Ferdinand de Saussure and Roland Barthes, as well as to the proponents of material culture and design history.[2] The earlier, diligent antiquarians such as Joseph Strutt, Eugène Viollet-le-Duc, and the later art historians, such as Otto von Falke, are also due credit for their empirically-based findings in the vast, opinionated, and by no means flawless compilations that helped establish the corpus of evidence.[3] It was clear to them that relevant research should be grounded in equal parts on close examination of surviving medieval artifacts *and* on their representation in the writings and images of the time.

From 1997 to the present, our Special Session on *Medieval Textiles and Dress: Object, Text, Image* at the International Congress on Medieval Studies at Western Michigan University in Kalamazoo has served as a forum for new work by scholars and practitioners in archaeology, art history, conservation, drama, economics, history, legal studies, literature, religion, and technology. The fourteen chapters that make up this book derive in part from the Kalamazoo sessions, while some other authors were invited to contribute. The volume also represents thematic diversity with a broad spatial and temporal span as it celebrates the encounter with attire, dress, and textiles in medieval society.

Today, "costume history" is disappearing as a designation, and is being replaced by "history of dress," and concepts such as "the study of culture of clothing and appearance" are current. This change, of course, is more than just a new rubric—the new emphasis, again borrowed from the literary theory of the 1970s—is implicit in Jauss's reception theory that calls for a removal of the "prejudices of historical objectivity."[4] This is an especially useful approach for the meager, usually decontextualized medieval material that we encounter in objects, texts, and images that once were executed for patrons and for purposes that were seldom recorded. Visual representations of dress and textile objects can never be taken at face value, of course, but the more challenging notion is to admit that we can probably never reconcile the medieval images of idealized dress, body, gesture, brilliance of color and pattern with any "accurate" idea of the medieval reality. After all, we probably also think that an attenuated, mannered female model drawn by Erté is more "authentically" representative of the 1920s than a photograph of chubby bathing beauties of the same time, or, indeed, their extant woolen swimsuits and caps.

One of the themes in this book concerns the phenomenon of medieval "fashion," both as it relates to clothing styles that appeared as novelties in works of art and in texts, and to the efforts of later writers who attempted to impose their ideas on the medieval evidence for purposes of identification and dating by fitting it into a linear history. No single definition of fashion is presented here, rather our authors recognize and work within established parameters and methodologies while proposing new interpretations (Blanc) and bringing attention to less accessible (L'Engle), lesser known (Heckett), and downright humble evidence to trace alternate paths of development (Effros, Williams), and advocating for their preservation (Crummy). Ideas derived from the meetings of cultures, and as results of political events in medieval society, are seen to influence "fashion" in significant ways (Heller, Jolly). It is especially the young noblemen who, unlike their female counterparts, could travel, go to war, and adopt foreign mores to create a "Youth Quake" on their return home (Blanc, Koslin). Some of the authors emphasize that fashion existed prior to the late medieval, explicitly rendered images of verisimilitude that many historians have equated with the "birth of fashion" (Effros, Snyder, Williams). Research in iconography has also broadened to include dress as carrier of meaning beyond the three estates, capable of subtle nuance when "read" with textual concordance (Anderson, Cottrell, Gilmore, Heller, Yanson).

To a remarkable extent, the research published here has been informed by paying attention to the smallest details in dress accessories including headdresses, hairstyles, closures, jewelry, patterns, hems, and girdles, and so on. (Anderson, Blanc, Cottrell, Crummy, Effros, Heckett, Jolly, Koslin, L'Engle, Snyder, Williams). Their reconstruction and reasons for being have been considered, as in material culture theory, from "inside-out," deriving the context from the objects rather than merely contextualizing them. The medieval awareness of and proclivity for archaizing elements, thought of today as "postmodern," have also been noted in the medieval corpus (Anderson, Heckett, Koslin, Snyder, Yanson). These successfully developed ideas bode well for the continued discourse on appearance, its accoutrements, and literal as well as metaphorical construction that can be observed in medieval dress and textiles.

Notes

1. From 1979, *Block* (Middlesex Polytechnic) has featured articles on art, design, and culture, and since 1997, *Fashion Theory: The Journal of Dress, Body & Culture,* ed. Valerie Steele, is published.

2. For the structuralist approach, see Ferdinand de Saussure, *A Course in General Linguistics* (London: Peter Owen, 1960); and Roland Barthes, *Mythologies* (London: Paladin, 1973); and his *Système de la Mode* (1967) in trans. as *The Fashion System* (New York: Hill and Wang, 1983). The study of material culture, using a progression of description, deduction, and speculation, is defined by Jules Prown in his "Mind in Matter: An Introduction to Material Culture Theory and Method," *The Winterthur Portfolio* (17, 1982). Design history was recently introduced in academic curricula; see, for an introduction John A. Walker, *Design History and the History of Design* (London: Pluto Press, 1989); and Adrian Forty, *Objects of Desire: Design and Society since 1750* (London, Thames and Hudson, 1986).

3. Joseph Strutt's 1796 *Complete View of the Dress and Habits of the People of England* is accessible in the 1970 facsimile of the 1842 reprint edited by J. R. Planché (London: Tabard Press, 1970); Eugène Viollet-le-Duc, *Dictionnaire raisonné du mobilier français de l'époque carolingienne à la renaissance,* vols. III-IV (Paris: Morel et Cie., 1872); and for textiles especially, Otto von Falke, *Kunstgeschichte der Seidenweberei* (Berlin: Ernst Wasmuth, 1921 (1913).

4. See Hans Robert Jauss, "Literary History as a Challenge to Literary Theory," in *New Directions in Literary History,* ed. Ralph Cohen (Baltimore: Johns Hopkins University Press, 1974), 13.

PART ONE

THE EARLY MIDDLE AGES

CHAPTER 1

APPEARANCE AND IDEOLOGY:
CREATING DISTINCTIONS BETWEEN CLERICS
AND LAYPERSONS IN EARLY MEDIEVAL GAUL

Bonnie Effros

The Symbolic Significance of Clerical Dress
in Early Medieval Legislation

Clothing and other sorts of bodily adornment represent an important means by which individuals and groups express their identity. Garments, hairstyles, and the display of related possessions may reveal the nature of a person's relationship to his or her contemporaries, whether with respect to religious belief, gender, age, ethnic affiliation, status, or membership in a kin group. These practices are culturally specific, but may share certain features with more distant communities through trade, political or military contact, or the desire to emulate powerful or ancient models. Mores regarding personal appearance thus often serve to reinforce the social order, just as their violation may challenge the status quo. As a consequence, the existence of specific sorts of dress to distinguish particular social groups may indicate important hierarchical divisions existing within a society that are not tied solely to legal rank or social status. Keeping in mind that laws prescribing personal appearance preserve the ideals of a regulating body rather than reflect what actually occurred in that kingdom or religious community, evidence for clothing nonetheless allows scholars to understand how nuances in the expression of difference in historical and contemporary communities are valued by their leaders.

Particularly interesting examples of these processes were attempts by clerics in the Merovingian period to distinguish their numbers from the lay nobility. These early efforts were of great importance, since beyond their consecration or taking of vows, the clergy were not separated from noble laypersons by more tangible factors such as status, ethnicity, or gender. After all, they originated from the same families as their lay competitors, had benefited from comparable political and economic connections, and were accustomed to a similar way of life during at least part of their childhood. Their authority thus depended in part upon their success at staking out an exclusive identity as church leaders, through not only their behavior but also their appearance. Emphasis on such differences likewise impeded clerics from returning to activities engaged in by the lay nobility but deemed inappropriate once they joined clerical ranks or monastic communities. Although not as comprehensive as those of the Carolingian era, Merovingian-period strategies for creating distinctions between clerics and laypersons through personal adornment provided important precedents for these later developments.

In early medieval Gaul, modest clothing and tonsure represented the primary means by which to distinguish clerics visibly from their lay contemporaries. Our most direct sources for these regulations survive in the canons of church councils and monastic *Rules*. Concern about the appearance of religious leaders constituted a repeated theme in ecclesiastical synods south of the Loire from the late fifth to seventh centuries. As early as circa 475, the *Statuta ecclesiae antiqua* stipulated in two measures that clerics not grow their hair, shave their beards, nor wear inappropriate clothing or shoes.[1] In 506, the twentieth canon of the Council of Agde further insisted upon clerical tonsure and suitable liturgical dress.[2] In order to make the visual and physical separation between priests and laymen legally binding, the fifth canon of the First Council of Mâcon (581–583) forbade clerics from wearing secular clothing and shoes; those found dressed improperly or carrying weaponry were to be sentenced to 30 days in isolation on bread and water.[3] Nearly a century later, the Council of Bordeaux (662–675) threatened punishment for clerics who bore lances or other weaponry or wore secular clothing.[4]

Although relatively infrequent, warnings of penalties for noncompliance indicated the continuing temptation for religious authorities to dress as laypersons as well as to possess armament. Canonical texts nonetheless did not yet closely define the exact nature of their ceremonial garments. Liturgical dress, including the requirement of a belt (*cingulum*) to tie the tunic of priests based on the precedent of Peter (John 21, 18), did not receive attention in Gaul before the eighth century. Surviving examples of leather belts with elaborate buckles from early medieval Gaul, such as the one alleged to have belonged to Caesarius of Arles found at Saint-Trophîme in Arles, however, may have been used in a religious context.[5] While it is possible that belts were so common that contemporaries did not consider it necessary to mention them, no extant evidence suggests that the *cingulum* was mandatory for clerics.

From early in the history of Christianity in Gaul, some clerics, and most famously Bishop Martin of Tours, were identified with military activities before their consecration.[6] Reform-minded bishops nevertheless periodically sought to bring about greater separation between professed religious leaders and their lay contemporaries. With the advent of clerical tonsure in Rome in the sixth century and its widespread establishment in Gaul in the seventh century, recognition of the benefits of establishing visual differences became even more firmly entrenched.[7] In an injunction dated to 589, participants at the Council of Narbonne decreed purple clothing an inappropriately worldly fashion for clerics. Noting that the color was unworthy of a priest's merit (*dignitas*), the canon stated that: "just as devotion is in [the cleric's] mind, likewise it should also be displayed on his body."[8] As purple was associated with imperial dignity, this measure was not simply an aesthetic issue. Rather, this regulation established a clear demarcation between the humility expected of clerics stemming from their devotion to God and the markers of high status normally conferred by men upon each other.[9] Further distinctions between the appearance of clerics and the lay aristocracy were left largely untouched in legislation prior to Carolingian reforming efforts.

Influenced by the *Rule of St. Benedict,* sixth-century clerical authorities also took steps to prevent monks from dressing as their secular counterparts and participating in activities deemed inappropriate to their vocation. Just as new converts (*conversi*) and penitents were required to rid themselves of personal possessions before being accepted into Christian congregations,[10] men who took monastic vows were obliged to give up worldly clothing. In exchange, they received a cowl, tunic, belt, handkerchief, sandals, and shoes from the abbot.[11] In the second book of his influential *Dialogues* dedicated to the life of Benedict of Nursia, Gregory I noted that monks could not carry handkerchiefs, even when they were received as gifts from nuns.[12] A variety of monastic *Rules* circulating in Gaul such as Aurelian's *Rule for Monks* (548) specified that the garments of monks were to remain undyed or be restricted to unimposing colors such as milk white and natural black.[13] Measures in Ferreolus's mid-sixth-century *Rule for Monks* likewise stipulated that the brethren were to be limited to a single change of modest clothing.[14] Having accepted a gift of clothing from a dying man, the Irish monk Fursey's soul was therefore punished with fire in a vision after his arrival in Gaul.[15] These texts did not so much describe monastic garments in a definitive manner as they set out to demarcate the appearance of monks and nuns from laypersons.

Although monks were restricted to plain-colored clothing, prohibited from wearing luxury fabrics, and obliged to be tonsured for reasons of humility, we must recognize that their robes had formerly signaled status. As opposed to short workmen's tunics or barbarian-inspired trousers, monks' robes alluded to their dignity and exclusivity, having been modeled after garments of the late Roman senatorial class,[16] or possibly after those of wandering Cynics in the East.[17] At the time of entering into the monastic house, ritual ceremonial exchanges of garments took place. These rites greatly heightened the value of what would have otherwise been modest clothing if acquired in ordinary circumstances.[18] Nonetheless, even with these added distinctions, the need to reiterate the separation between lay nobles and monastic inmates remained. The *Regula cuiusdam patris ad monachis* forbade travel by horse and chariot to all brothers except those who were ill; it penalized the disobedient with separation from the rest of the community.[19] Entailing not only a restriction of the monks' freedom of movement, this measure also deprived them of prerogatives indicative of a noble lifestyle. These customs hindered extended contact between monks and lay contemporaries. The attraction of the secular world must have been particularly strong in urban centers such as Arles, where authorities may have permitted chariot races as late as the sixth century.[20]

Nuns faced even harsher limitations and penalties for disobedience with respect to their appearance upon entry into the cloister. The *Rule* of Caesarius of Arles, written for his sister Caesaria's congregation early in the sixth century, represents the first known document in Gaul to advocate permanent claustration for nuns.[21] Used at a number of houses, including Radegund's monastery of the Holy Cross in Poitiers after 567,[22] Caesarius's *Rule* legislated that female religious were to adopt the monastic habit in conjunction with consecration. Prior to taking vows, nuns were expected to give away or sell their possessions.[23] Their monastic gowns were to be simple, modest in color, and could not be considered their own.[24] Not only did chapter 56 of the *Rule* stipulate that nuns had to wear clothing and veils appropriate to their office and bind their hair,[25] but they had to weave and sew their own garments as well:

> They [the nuns' garments] shall be made in the monastery through the diligence of the prioress and the careful attention of the sister in charge of wool work, and distributed by the mother of the monastery to each according to her reasonable necessities. There should be no dyeing done in the monastery, except, as is stated above, in plain or milk white color [*laia vel lactina*] because other colors do not befit the humility of a virgin.[26]

Weaving constituted a symbol of the sisters' humility, chastity, and even charismatic authority.[27]

The strict standards of Caesarius's *Rule* were by no means unique to the early medieval West with respect to restrictions on clothing. The anonymous seventh- or eighth-century *Rule* often referred to as the *Regula cuiusdam patris ad virgines* insisted that the nuns avoid vice through the disposal of all of their former possessions, including clothing and shoes, prior to taking their vows.[28] Yet, such *Rules* also had the intention of relieving nuns of continuing obligations to their families and superiors. Caesarius's monastic *Rule* for nuns forbade them from taking in clothing for washing, sewing, storing, or dyeing, from anyone outside the monastery without the abbess's permission.[29] Regulation of nuns' activities and physical appearance, especially through veiling, further reinforced the objectives of their permanent separation from lay life within the walls of the monastery.[30] Contemporary saints' *Lives* also celebrated the merits of these formal prescriptions with examples of the sanctity of nuns who renounced all but the most simple items of dress. For instance, in the early sixth-century *Vita Genovefae,* the virgin, who was not cloistered, received a copper necklace inscribed with a cross from Germanus of Auxerre, who thereby instructed her to avoid worldly riches.[31] Austreberta's eighth-century hagiographer noted that she always dressed very modestly.[32]

The poet and cleric Venantius Fortunatus's dramatic account of a Merovingian queen's decision to enter into the religious life in his *Vita sanctae Radegundis,* written shortly after her death in 587, illustrated the way in which clothing marked Radegund's transition to a religious existence. After Radegund fled from her husband, the king, Chlotar I, who had recently murdered her brother, she sought permanent freedom from him by taking vows as a nun. Bishop Medard of Noyon, fearing the wrath of the king if he were to permit his wife to swear lifelong chastity, hesitated to consecrate Radegund as a nun. She subsequently presented herself to him in changed garb and thereby persuaded him to ordain her as a deaconess, a less radical path than taking monastic vows.[33] At this time, she offered silk garments and bejeweled possessions including a diadem on the altar. She also distributed pieces of her golden girdle as alms among the poor. Although Radegund was not known for her strict adherence to the *Rule,* her exchange of royal clothing for more modest dress signaled her irreversible entry into the religious life.[34] While her outward adornment mirrored her spiritual transformation, however, her role as an astute player in Merovingian politics did not end at this time.[35]

In reality, it is abundantly clear that obedience to strict monastic regulations regarding clothing did not always prevail. Imitation of the fashions of lay contemporaries and retention of certain luxuries must have been

common among members of monastic communities. Since it is largely
prescriptive written evidence upon which historians must rely, however,
the range of compliance to monastic legislation is difficult to measure. Enforcement of the *Rule* must have varied significantly from house to house,
something that is seldom reflected in hagiographical accounts. Only in
very rare situations is it possible to juxtapose textual sources with the burial remains of a monk or nun. The majority of surviving clothing from
these few examples, moreover, did not reflect typical instances of monastic
interment but rather the possessions and bones of saints preserved as relics.
The results of such comparisons are nonetheless thought-provoking. The
most well-known relic of clothing is believed to have belonged to the
Neustrian queen Balthild, who lived the last years before her death in
680/681 at her foundation of Chelles. The garment comprised an embroidered blouse depicting a pectoral cross and multiple necklaces. The imagery may be alternately viewed as a sign of her humility (in not utilizing
real jewels) or Balthild's attempt to imitate Byzantine imperial dress following her fall from power.[36] She may have worn this garment during her
lifetime, but its intact state indicates that she was never buried in it.[37]
Balthild herself was interred dressed in a large semicircular cloak of red
color with yellow fringes with a brooch at her chest; her plaited hair was
wrapped in gold and silk bands dyed in the colors of red, yellow, and green.
The first abbess of Chelles, Bertilla, a former nun of Jouarre who died
some time after 704, was buried in a brown silk tunic with yellow stripes
and edging.[38] None of these garments reflected the modest standards advocated by contemporary monastic legislation.

Balthild's and Bertilla's relics are admittedly far from representative of the
monastic experience of early medieval nuns of noble parentage. Balthild,
having lost the throne, did not choose to spend her final years at Chelles but
likely went there under duress.[39] Because their remains were considered holy
by their successors, moreover, neither woman's burial can be assumed to
document typical monastic attire in life or death.[40] Yet, while prudence demands recognition that luxury objects may have been used to show respect
for the perceived sanctity of the abbey's royal benefactor and first abbess, the
possibility exists that monastic women did not always adhere to the strict
provisions of the *Rule* regarding clothing. If nuns were concerned with pro-
jecting the level of humility achieved during his or her lifetime, including
the way in which they had been clothed, surely one would expect their fu-
nerals to have reflected this desire for modesty. At the very least, these burials indicate that lavish display must not have offended the brethren or sisters
honoring the relics of former inhabitants of their houses.[41]

For comparative purposes, it is useful to note that contemporary Anglo-
Saxon texts such as Aldhelm's (d. 709) *De virginitate* condemned abuses of

prescribed customs of personal adornment among both male and female noble members of monasteries in England. Acting in direct defiance of canonical restrictions, monks and nuns apparently wore luxury items after taking vows, including underclothing of linen, red and blue tunics, head coverings and sleeves with silk borders, shoes of red-colored leather, and white and colored veils.[42] Not only did such garments contradict insular ascetic practices, but they also blurred visual and symbolic distinctions between members of monastic houses and lay nobility. Aldhelm's condemnation of excess reinforced the belief that silk cloth, just as gold and silver, was considered a valuable most appropriately kept in a king's treasury or taken as plunder.[43]

Regulating Bodily Adornment among the Laity

Not surprisingly, far fewer prescriptions in Merovingian-period Gaul addressed the appearance of the lay nobility. The most famous effort to enforce social distinctions among inhabitants via personal adornment was the legendary Frankish restriction of long hair to members of the royal dynasty, such as recorded by Gregory of Tours.[44] Although no surviving code attests to this custom, and it is unlikely that Merovingian kings were the only males with long hair, historical accounts indicate that the gesture of cutting kings' hair was by no means merely a symbolic form of public humiliation. Monarchs who were overturned, if lucky enough not to be assassinated, received haircuts or monastic tonsure.[45] The objective of tonsuring a king theoretically meant that if he sought again to take up arms, regain the throne and defiantly grow out his hair, he broke with his consecration or monastic profession and technically forfeited the throne through excommunication.[46] The effectiveness of such measures was decidedly mixed and did not thwart Dagobert II's effort to assume the throne of Austrasia 20 years after his tonsure and banishment by Grimoald, mayor of the palace, circa 656.[47] In addition, the symbolism of short hair was not always negative. Pepin the Short's adoption in 737 by the Lombard king Liutprand was marked by a ritual haircut.[48]

Beyond noting the king's hairstyle, Gregory of Tours observed that the purple robes, military belt, and diadem of Clovis distinguished him from contemporary nobles.[49] Although sixth-century accounts were likely somewhat anachronistic as to the dress of earlier Merovingian kings, they signaled what garments were perceived as regularly associated with monarchs, even if these items were not exclusively royal in their use. While leaders of the Franks such as Childeric I and the individual found bearing what might have been the ring of queen Aregund were buried with signet rings, these were not their sole possible application.[50] Others certainly had access to signet and name rings.[51] Grooming, dress, and the treasuries of kings and queens, rather than the more formal system of royal insignia developed during the Carolingian period, expressed their unique status in life and death.[52]

Evidence is scarce for secular or religious legislation affecting the manner in which the more general population dressed. The Frankish *leges* did not regulate social status or ethnic identity on the basis of personal appearance. Likewise, the primary concern of clerical legislation was liturgical clothing; a few exceptional measures, however, addressed the bodily appearance of Christians other than clerics. An unusual prescription appeared in the seventh-century *Canones Wallici,* written circa 550–650 in rural Wales, but which assumed their final form in Brittany. The collection forbade Catholics from wearing their hair in the "barbarian fashion" under threat of excommunication.[53] Hayo Vierck has also argued that Christians were probably also distinguished from pagans by voluntary differences in their clothing.[54] Since written evidence for such customs is largely nonexistent, however, modern attitudes toward the expression of religious identity should not be read into material artifacts.[55]

More typical legislation prescribing practices regarding appearance included canons reinforcing distinctions between ordinary Christians and those undergoing self-imposed penance or baptism. Clerics at the Council of Agde (506) warned that members of penitential orders who refused to shave their heads or adopt the customary hair shirt would be excommunicated.[56] More general descriptions in the *Liber historiae Francorum* (727) and the works of Gregory of Tours revealed that newly baptized were clothed in white and mourners dressed in black.[57] Local churches no doubt regulated the former on an informal basis, whereas they did not have the same authority over the latter. The modes of dress most often addressed in canon law were those that prevented clerics from relaxing the distinctions characteristic of their way of life. Restrictions on the appearance, garments, and personal possessions of the majority of the laity, by contrast, did not belong to the domain of canon law, as was more often the case in Ireland.[58]

Nor does archaeological evidence help greatly to identify the customs regulating personal appearance of particular religious or lay groups in Merovingian Gaul. Grave goods do not so much reveal customary choices of dress among the living as point to communal and familial funerary traditions. Moreover, because most excavated finds of clothing result from their deposition with the deceased whose identity is unknown, drawing conclusions from such contexts regarding the mores of clerics versus laypersons is difficult. Artifactual remains are not conclusive in determining the rank or status of deceased individuals, despite efforts, for instance, to demonstrate links between weapon sets and the legal standing of those with whom they were buried.[59] Similar problems, for that matter, affect our ability to measure the impact of ethnicity on the wearing of weaponry and brooches.[60] The influence of religious authority on burial adornment is even more difficult to ascertain.[61] Rank, social status, religious belief, and ethnicity were not communicated via legally-sanctioned modes of dress, but rather through the innovation of elites in choosing garments and other sorts of adornment, and others' imitation of these leaders within an accepted social framework. The quality, quantity, and choice of garments and personal possessions to which kin had access on the basis of wealth and membership in the circles in which they were exchanged, influenced the expression of a person's identity.[62] Families' donations of various symbolic artifacts to graves represented a form of ritual exchange forging or renewing the relationships necessary for the commemoration of the deceased and their own survival following such a loss.[63]

Clothing and the Law

Keeping in mind the silence of Merovingian-period clerical and lay authorities with respect to bodily adornment, what other concerns may be identified in the written sources with regard to personal appearance? References to garments, jewelry, and weaponry in the *leges* in early medieval Gaul consist of a small number of statutes regulating the inheritance of such objects.[64] Laws punishing grave theft also mentioned specific possessions such as women's bracelets in some circumstances.[65] Although beyond the scope of the present discussion, the early ninth-century *Lex Thuringorum* 38 stipulated, for instance, that the perpetrator of the robbery of female adornment (*ornamenta muliebra* or *rhedo*) recompense the victim threefold and pay a penalty of 12 *solidi* to public authorities.[66] If bodily adornment were truly regulated on the basis of rank, these statutes would have surely been included among measures devoted to questions of legal standing and the respective rights and penalties accorded to free and non-free, male and female, and children and adult members of early medieval communities.[67] For the most part, law codes operating in Gaul and outlying regions concentrated instead on protecting the transmission of clothing and weaponry rather than restricting certain possessions to specific groups.

Sixth- and seventh-century society in Gaul remained too fluid for social distinctions based on bodily adornment to become firmly established.[68] Status constituted just one aspect of the deceased's identity that might be expressed through personal appearance in life and at burial, but it was primarily financial resources and participation in networks of gift exchange and foreign trade that enabled or hindered access to luxury items and other possessions. As made evident by the wide array of grave goods, including a brooch produced in a workshop in a Frankish region but found in a richly endowed female sepulcher at Wittislingen near the Danube,[69] visual demarcations indicative of distinctions among the laity reflected the networks to which families had access. No legislation, by comparison, insisted that individuals born in a particular region retain their native dress throughout their lives.[70] Consequently, secular measures contrasted sharply with the religious canons directed toward the reinforcement of divisions within and outside of the religious hierarchy because the latter were not self-regulating. As members of the nobility, clerics had access to luxury garments and other objects of adornment and therefore had to be discouraged from using them with legislation designed to reinforce their differences from the laity.

For both temporal and religious leaders, clothing and related possessions represented objects of significant value because of their inherent worth as well as their ability to convey symbolic meaning. Yet, their attitudes toward these possessions varied as a result of how they were to be utilized. In the Frankish *leges,* garments constituted a category of property viewed far differently than by early twentieth-century German legal historians who argued for the alleged existence of the customs of *Heergewäte* and *Gerade,* by which the Germanic peoples buried their dead with personal possessions because they could not be bequeathed to descendants.[71] Among early medieval laypersons, few restrictions affected the disposal of personal items. No written evidence exists for efforts to regulate adornment with respect to status or ethnicity, since secular law promulgated in the Merovingian kingdoms primarily addressed inheritance and theft. Among clerics, by contrast, canonical measures focused on preventing their numbers from interacting too often or mingling too closely with their lay contemporaries. Maintaining visual differences was critical to heightening the status and preserving the identity of clerics who could be distinguished in few other ways from the lay elite families in which they had been raised.[72]

Notes

1. Canon 25: "Clericus nec comam nutriat nec barbam radat." Canon 26: "Clericus professionem suam etiam habitu et incessu probet et ideo nec vestibus nec calceamentis decorem quaerat." C. Munier, ed., *Concilia Galliae A.314-A.506, Corpus christianorum* series latina [*CCSL*]148 (Turnhout: Typographi Brepols editores Pontificii, 1963), 171.

2. Munier, ed., *Concilia Galliae, CCSL* 148, 202.

3. "Ut nullus clericus sagum aut vestimenta vel calciamenta saecularia, nisi quae religionem deceant, induere praesumat. Quod si post hanc defini-

tionem clericus aut cum indecenti veste aut cum arma inventus fuerit, a seniorebus ita coherceatur, ut triginta dierum conclusione detentus aquam tantum et modeci panis usu diebus singolis sustentetur." Charles de Clercq, ed., *Concilia Galliae A.511-A.695, CCSL* 148A (Turnhout: Typographi Brepols editores Pontificii, 1963), 224.

4. *Conc. Burdigalense* canon 1: "Ut abitum concessum clerici religiose habitare debeant et nec lanceas nec alia arma nec vestimenta secularia habere nec portare debeant . . . ut, qui post hanc definitionem hoc agere aut adtemtare presumerit, canonica feriatur sententia." De Clercq, ed., *Concilia Galliae, CCSL* 148A, 312. See also *Conc. Latunense* (now Saint-Jean-de-Losne, Côte d'Or) (673–5) canon 2: "Ut nullus episcoporum seu clericorum arma more seculario ferre praesumat." De Clercq, ed., *Concilia Galliae, CCSL* 148A, 315.

5. Joseph Braun, *Die liturgische Gewandung im Occident und Orient: Nach Ursprung und Entwicklung, Verwendung und Symbolik*, reprint ed. (Darmstadt: Wissenschaftliche Buchgesellschaft, 1964), 101–16; H. Leclercq, "Ceinture," in *Dictionnaire d'archéologie chrétienne et de liturgie* 8 (Paris: Létouzey et Ané, Éditeurs, 1909), cols. 2779–94.

6. Raymond Van Dam, *Leadership and Community in Late Antique Gaul* (Berkeley: University of California Press, 1985), 124–34.

7. Lynda L. Coon, *Sacred Fictions: Holy Women and Hagiography in Late Antiquity* (Philadelphia: University of Pennsylvania Press, 1997), 62–3. In Ireland, by contrast, one measure advocated that clerics continue to wear shoes, perhaps indicating that some were pursing humility with what was considered an excess of zealous behavior. Hermann Wasserschleben, ed., *Die irische Kanonensammlung* 66, 8, 2d ed. (Leipzig: Verlag von Bernhard Tauchnitz, 1885), 237. No similar measure survives from Gaul.

8. "Hoc regulariter definitum est, ut nullus clericorum vestimenta purpurea induat, quae ad iactantiam pertinet mundialem, non ad religiosam dignitatem, ut sicut est devotio in mente, ita et ostendatur in corpore; quia pur[pur]a maxime laicorum potestate preditis debetur, non religiosis. Quo quisque non observaverit, ut transgressorem legis choercendum." De Clercq, ed., *Concilia Galliae, CCSL* 148A, 254.

9. Color also appears to have had great significance in relation to clothing in early medieval Ireland. The Law of Fosterage II.148–9 (*Senchus Mor*) specified the frequency and quantity of colored garments permitted to those of various social ranks. Arthur Haddan West and William Stubbs, eds., *Councils and Ecclesiastical Documents Relating to Great Britain and Ireland* 2, 2 (Oxford: Clarendon Press, 1878), 350–1. This code differed in a very basic way from the measures promulgated in the above mentioned synods in Gaul, since its main concern was to reinforce distinctions among laypersons based primarily on legal and economic criteria.

10. Cyril Vogel, "La discipline pénitentielle en Gaule des origines au IXe siècle: Le dossier hagiographique," *Revue des sciences religieuses* 30 (1956): 176–7.

11. "Et ut hoc vitium peculiaris radicitus amputetur, dentur ab abbate omnia quae sunt necessaria, id est cuculla, tunica, pedules, caligas, bracile, cultellum, graphium, acum, mappula, tabulas, ut omnis auferator necessitatis excusatio." Timothy Fry, ed., *The Rule of Saint Benedict* 55, 18 (Collegeville: The Liturgical Press, 1981), 262–5.

12. Gregory the Great, *Dialogues* 2.18.1–2, *Sources chrétiennes* [SC] 260, Adalbert de Vogüé, ed. and trans. (Paris: Les Éditions du CERF, 1979), 196–7.

13. "Vestimenta alio colore non induatis nisi laia, lactina, et nigra nativa." Aurelian, *Regula ad monachos* 26, in *Patrologia latina* [PL] 68, Jacques-Paul Migne, ed. (Paris: Apud editorem, 1847), col.391. By contrast, the fifth canon of the Irish *Rule of Ailbe of Emly* specifically forbade fringes of red leather as well as blue and red clothing to monks. Joseph O Neill, "The Rule of Ailbe of Emly," *Ériu* 3 (1907): 96–7.

14. "Vestimenta supersufficientia monachus non requirat: qui non amplius quam quod ad quotidianum usum abbas necessarium viderit, consequatur; dicente Domino ad discipulos:'Neque duas tunicas habeatis' [Luke 7]. . . ." Ferreolus of Uzés, *Regula ad monachos* 14, in *PL* 66, Jacques-Paul Migne, ed. (Paris: Apud editorem, 1847), col.964.

15. W.W. Heist, ed., *Vita S. Fursei* 22, in *Vitae sanctorum Hiberniae ex codice olim Salmanticensi nunc Bruxellensi,* Subsidia hagiographica 28 (Brussels: Société des Bollandistes, 1965), 47.

16. Barbara F. Harvey, *Monastic Dress in the Middle Ages: Precept and Practice* (Canterbury: The William Urry Memorial Trust, 1988), 10–12.

17. Jacques Dubois, "Utilisation religieuse du tissu," in *Tissu et vêtement: 5000 ans de savoir-faire, 26 avril–30 novembre 1986* (Guiry-en-Vezin: Musée archéologique départment du Val-d'Oise, 1986), 144–6.

18. Patrick J. Geary, "Sacred Commodities: The Circulation of Medieval Relics," in *The Social Life of Things: Commodities in Cultural Perspective,* Arjun Appadurai, ed. (Cambridge: Cambridge University Press, 1986), 172–4.

19. "Monachos in curribus et in equis discurrere, praeter infirmos, sed debiles et claudos, non permittimus. Si autem hoc fecerint, alieni sint ab unitate fratrum." Jacques-Paul Migne, ed., *Regula cuiusdam patris ad monachos* 21, in *PL* 66, col.992.

20. Yitzhak Hen, *Culture and Religion in Merovingian Gaul A.D. 481–751* (Leiden: E. J. Brill, 1995), 219–26.

21. On the origins of this *Rule,* see: Germain Morin, "Problèmes relatifs à la règle de S. Césaire d'Arles pour les moniales," *Revue Bénédictine* 44 (1932): 5–20; D.C. Lambot, "Le prototype des monastères cloîtrés de femmes: L'abbaye Saint-Jean d'Arles (VIᶜ siècle)," *Revue liturgique et monastique* 23 (1937–1938): 170–4.

22. Brian Brennan, "St. Radegund and the Early Development of Her Cult at Poitiers," *Journal of Religious History* 13 (1985): 342–3.

23. "Quae autem viduae, aut maritis relictis, aut mutatis vestibus ad monasterium veniunt, non excipiantur, nisi antea de omni facultaticula sua, cui voluerint, cartas, aut donationes, aut venditiones faciant, ita ut nihil suae

potestati, quod peculiariter aut ordinare aut possidere videantur, reservent. . . ." Caesarius of Arles, *Ad regulam virginum* 5, in *Sanctus Caesarius Arelatensis, Opera omnia* 2, Germain Morin, ed. (Bruges: Jos. van der Meersch, 1942), 103.

24. "Nemo sibi aliquid iudicet proprium, sive in vestimento, sive in quacumque alia re." Caesarius of Arles, *Ad regulam virginum* 17, in *Sanctus Caesarius Arelatensis, Opera omnia* 2, 105.

25. Conrad Leyser, "Long-Haired Kings and Short-Haired Nuns: Power and Gender in Merovingian Gaul," *Medieval World* 5 (1992): 37–42.

26. Caesarius of Arles, *Ad regulam virginum* 44, in *Sanctus Caesarius Arelatensis, Opera omnia* 2, C113–114. Translation adapted from Maria Caritas McCarthy, *The Rule for Nuns of Caesarius of Arles: A Translation with a Critical Introduction*, Catholic University of America: Studies in Mediaeval History, new ser. 16 (Washington: Catholic University of America Press, 1960), 44.

27. Bonnie Effros, "Symbolic Expressions of Sanctity: Gertrude of Nivelles in the Context of Merovingian Mortuary Custom," *Viator* 27 (1996): 1–5; Coon, *Sacred Fictions,* 41–4.

28. "Amputandum ergo est hoc vitium radicitus ab omni monacha, ut nullam rem vel in vestimentis, seu in calceamentis, vel in quibuslibet rebus sibi vindicet, vel suum esse dicat, nisi quantum ex abbatissae iussione penes se praecipitur retinere. . . ." Jacques-Paul Migne, ed., *Regula cuiusdam patris ad virgines* 17, in *PL* 88 (Paris: Apud editorem, 1850), col.1066; Felice Lifshitz, "Is Mother Superior? Towards a History of Feminine *Amtscharisma,*" in *Medieval Mothering,* John Carmi Parsons and Bonnie Wheeler, eds. (New York: Garland Publishing, Inc., 1996), 126.

29. "Nulla ex vobis extra iussionem abbatissae praesumat clericorum sive laicorum, nec parentum, nec cuiuscumque virorum sive mulierum extranearum vestimenta, aut ad lavandum, aut ad consuendum, aut ad reponendum, aut ad tingendum accipere sine iussione abbatissae. . . ." Caesarius of Arles, *Ad regulam virginum* 46, in *Sanctus Caesarius Arelatensis, Opera omnia* 2, 114.

30. Coon, *Sacred Fictions,* 33–41; Gabriella Schubert, *Kleidung als Zeichen: Kopfbedeckungen im Donau-Balkan-Raum.* (Wiesbaden: Harrassowitz Verlag, 1993), 137–42.

31. "Cui sanctus Germanus nummum aereum Dei notu allatum, habentem signum crucis, a tellure colligens pro magno munere dedit, inquiens ad eam: 'Hunc transforatum pro memoria mei et collo suspensum semper habeto; nullius mettali neque auri neque argenti seu quolibet margaritarum ornamento collum, saltim digitos tuos honerare paciaris." Bruno Krusch, ed., *Vita Genovefae virginis Parisiensis* 6, in *Monumenta Germaniae historica* [*MGH*]: *Scriptores rerum Merovingicarum* [*SRM*] 3 (Hanover: Impensis bibliopolii Hahniani, 1896), 217; Coon, *Sacred Fictions,* 25–6.

32. Ioannes Bollandus and Godefridus Henschenius, eds., *Vita S. Austrebertae virginis* 13, in *Acta sanctorum* [*AASS*] Februarius II (Antwerp: Apud Iacobum Meursium, 1663), 422.

33. Venantius Fortunatus, *De vita sanctae Radegundis* 1,12, in *MGH: SRM* 2, new ed., Bruno Krusch, ed. (Hanover: Impensis bibliopolii Hahniani, 1956), 368.

34. "Mox indumentum nobile, quo celeberrima die solebat, pompa comitante, regina procedere, exuta ponit in altare et blattis, gemmis ornamentis mensam divinae gloriae tot donis onerat per honorem. Cingulum auri ponderatum fractum dat opus in pauperum. Similiter accedens ad cellam Sancti Iumeris die uno, quo se ornabat felix regina, conposito, sermone ut loquar barbaro, stapione, camisas, manicas, cofias, fibulas, cuncta auro, quaedam gemmis exornata per circulum, sibi profutura sancto tradit altario."Venantius Fortunatus, *De vita sanctae Radegundis* 1,13, in *MGH: SRM* 2, p.369; Bonnie Effros, "Images of Sanctity: Contrasting Descriptions of Radegund by Venantius Fortunatus and Gregory of Tours," *UCLA Historical Journal* 10 (1990): 46–7.

35. Georg Scheibelreiter, "Königstöchter im Kloster: Radegund (+587) und der Nonnenaufstand von Poitiers (589)," *Mitteilungen des Instituts für Österreichische Geschichtsforschung* 87 (1979): 1–5; 10; Brennan, "St. Radegund," 340–6.

36. H. E. F. Vierck, "La 'chemise de Sainte-Balthilde' à Chelles et l'influence byzantine sur l'art de cour mérovingien au VIIe siècle," in *Actes du colloque international d'archéologie: Centenaire de l'abbé Cochet* (Rouen: Musée départmental des antiquités de Seine-Maritime, 1978), 539ff.

37. I thank Alain Dierkens for his personal communication on this topic.

38. Jean-Pierre Laporte and Raymond Boyer, *Trésors de Chelles: Sépultures et reliques de la reine Bathilde (+ vers 680) et de l'abbesse Bertille (+ vers 704),* Catalogue de l'exposition organisée au Musée Alfred Bonno (Chelles: Société archéologique et historique, 1991), 22–34.

39. Richard A. Gerberding, *The Rise of the Carolingians and the 'Liber historiae Francorum'* (Oxford: Clarendon Press, 1987), 68–9.

40. On the bejeweled tomb of Eligius, bishop of Noyon, as a center of religious veneration, see: Peter Brown, *The Rise of Western Christendom: Triumph and Diversity AD 200–1200* (Oxford: Blackwell Publishers, 1996), 104.

41. Bonnie Effros, *Caring for Body and Soul: Burial and the Afterlife in the Merovingian World* (University Park: Pennsylvania State University Press, 2002), 20–3.

42. Aldhelm, *De virginitate (prosa)* 58, in *MGH: Auctores antiquissimi [AA]* 15, Rudolf Ewald, ed. (Berlin: Apud Weidmannos, 1919), 317–18; Hayo Vierck, "Zur angelsächsischen Frauentracht," in *Sachsen und Angelsachsen: Ausstellung des Helmsmuseums, Hamburgisches Museum für Vor- und Frühgeschichte 18. November 1978 bis 28. Februar 1979* (Hamburg: 1978), 256.

43. Matthias Hardt, "Royal Treasures and Representation in the Early Middle Ages," in *Strategies of Distinction: The Construction of Ethnic Communities 300–700,* Transformation of the Roman World [TRW] 2, Walter Pohl and Helmut Reimitz, eds. (Leiden: E. J. Brill, 1998), 266; 271.

44. Gregory of Tours, *Libri historiarum X* 2,9, rev. ed., *MGH: SRM* 1, 1, Bruno Krusch, ed., (Hanover: Impensis bibliopolii Hahniani, 1951), 57; Averil Cameron, "How Did the Merovingians Wear Their Hair?" *Revue belge de philologie et d'histoire* 43 (1965): 1203–16.

45. Leyser, "Long-Haired Kings," 37–42.

46. Margarete Weidemann, *Kulturgeschichte der Merowingerzeit nach den Werken Gregor von Tours*, Römisch-Germanisches Zentralmuseum Monographien 3 (Mainz: Verlag des Römisch-Germanischen Zentralmuseums, 1982), 1: 20–3; 323–4; 2: 213.

47. Bruno Krusch, ed., *Liber historiae Francorum* 43, in *MGH: SRM* 2 (Hanover: Impensis bibliopolii Hahniani, 1888), 316; Eddius Stephanus, *Vita Wilfridi I episcopi Eboracensis* 28, Wilhelm Levison, ed., in *MGH: SRM* 6 (Hanover: Impensis bibliopolii Hahniani, 1913), 221–2; Gerberding, *The Rise of the Carolingians*, 47–8; 71–2.

48. Jörg Jarnut, "Die Adoption Pippins durch König Liutprand und die Italienpolitik Karl Martells," in *Karl Martell in seiner Zeit*, Jörg Jarnut, Ulrich Nonn, and Michael Richter, eds., Beihefte der Francia 37 (Sigmaringen: Jan Thorbecke Verlag, 1994), 217–8. On the *barbaratoria*, see: Hen, *Culture and Religion*, 137–43.

49. With reference to Clovis's appointment as consul by the emperor, Gregory of Tours observed: "Igitur ab Anastasio imperatore codecillos de consolato accepit, et in basilica beati Martini tunica blattea indutus et clamide, inponens vertice diademam." Gregory of Tours, *Libri historiarum X* 2, 38, *MGH: SRM* 1,1, 88–9. As for queens, see Venantius Fortunatus's account above of Radegund's charitable donation of her personal possessions.

50. Jean-Jacques Chiflet, *Anastasis Childerici I. Francorum regis sive thesaurus sepulchralis* (Antwerp: Ex officina Plantiniana Balthasaris Moreti, 1655), 128; Michel Fleury, "Le monogramme de l'anneau d'Aregonde," *Les Dossiers de Archéologie* 32 (1979): 43–5; Patrick Périn, "A propos de la datation et de l'interprétation de la tombe n.49 de la basilique de Saint-Denis, attribuée à la reine Arégonde, épouse de Clotaire Ier," in *L'art des invasions en Hongrie et en Wallonie: Actes du colloque tenu au Musée royal de Mariemont du 9 au 11 avril 1979*, Monographies du Musée royal de Mariemont [MRM], 6 (Morlanwelz: MRM, 1991), 11–30.

51. M. M. Deloche, *Étude historique et archéologique sur les anneaux sigillaires et autres des premiers siècles du moyen âge* (Paris: Ernest Leroux, Éditeur, 1900).

52. "Tunc egressus Chilpericus a Turnaco cum uxore sua ac populo, vestitum Sighiberto vestibus ornatis apud Lambrus vicum sepelivit." Krusch, ed., *Liber historiae Francorum* 32, in *MGH: SRM* 2, 296–7. "Mallulfus itaque Silvanectinsis episcopus, qui in ipso palatio tunc aderat, indutumque eum [Chilpericum] vestibus regalibus, in nave (in villa quae dicitur Calla) levato, cum hymnis et psallentio cum Fredegunde regina vel reliquo exercitu Parisius civitate in basilica beati Vincenti martyris eum sepelierunt." Krusch, ed., *Liber historiae Francorum* 35, in *MGH: SRM* 2, 304; Alain Erlande-Brandenburg, *Le roi est mort: Étude sur les funérailles, les sépultures et les*

tombeaux des rois de France jusqu'à la fin du XIIIᵉ siècle, Bibliothèque de la Société française d'archéologie 7 (Geneva: Droz, 1975), 7; Hardt, "Royal Treasures," 256–65.

53. Canon 61: "Si quis catholicus capillos promisserit more barbarorum, ab aeclesia Dei alienus habeatur et ab omni Christianorum mensa, donec delictum emendat." Ludwig Bieler, ed., *The Irish Penitentials* (Dublin: The Dublin Institute for Advanced Studies, 1963), 7; 148.

54. Hayo Vierck, "Religion, Rang und Herrschaft im Spiegel der Tracht," in *Sachsen und Angelsachsen,* 272.

55. See the short list of excerpts compiled by Weidemann from the writings of Gregory of Tours with respect to the linen and silk clothing of the nobility, as well as the appearance their hair and beards in various circumstances. Weidemann, *Kulturgeschichte* 2, 362–5.

56. *Conc. Agathense* canon 15: "Paenitentes, tempore quo paenitentiam petunt, impositionem manuum et cilicium super caput a sacerdote sicut ubique constitutum est, consequantur; et si aut comas non deposuerint, aut vestimenta non mutaverint, abiiciantur et nisi digne paenituerint, non recipiantur." Munier, ed., *Concilia Galliae,* CCSL 148, 201; Cyril Vogel, *La discipline pénitentielle en Gaule des origines à la fin du VII siècle,* Thèse pour le doctorat en théologie, Strasbourg 1950 (Paris: Letouzey et Ané, 1952), 104–6.

57. Gregory of Tours, *Libri historiarum X* 5, 11, *MGH: SRM* 1, 1, 206. "Magnus quoque planctus hic omni populo fuit; nam mulieres cum viris suis lugentes flebant, nigris vestibus indutae, percussa pectora, hoc funus [Chlodeberti] sunt prosequutae." Krusch, ed., *Liber historiae Francorum* 34, in *MGH: SRM* 2, 301. On baptismal white, refer also to: Edmond Le Blant, *Inscriptions chrétiennes de la Gaule antérieures au VIIIᵉ siècle* 1 (Paris: A l'Imprimerie impériale, 1856), 476–7; Dubois, "Utilisation religieuse," 148.

58. The so-called *Poenitentiale Cummeani* 14, 9, directed that: "Mulieres possunt sub nigro velamine accipere sacrificium." Hermann Joseph Schmitz, ed., *Die Bussbücher und die Bussdisciplin der Kirche* 1 (Mainz: Verlag von Franz Kirchheim, 1883), 643. A chapter of the Irish canonical collection advocated modest dress for Christians on the authority of Jerome. Wasserschleben, ed., *Die irische Kanonensammlung* 46, 1, 235.

59. But see: Joachim Werner, "Zur Entstehung der Reihengräberzivilisation: Ein Beitrag zur Methode der frühgeschichtlichen Archäologie," *Archaeologia Geographica* 1 (1950): 25; Heiko Steuer, "Zur Bewaffnung und Sozialstruktur der Merowinger," *Nachrichten aus Niedersachsens Urgeschichte* 37 (1968): 30–74.

60. Pohl, "Telling the Difference," 32–40; Bonnie Effros, "Dressing Conservatively: A Critique of Recent Archaeological Discussions of Women's Brooches as Markers of Ethnic Identity," in *Gender and the Transformation of the Roman World: Women, Men and Eunuchs in Late Antiquity and After, 300–900 CE,* Julia Smith and Leslie Brubaker, eds. (Cambridge: Cambridge University Press, in press). But, in favor of the strict correspondence

between women's clothing and ethnicity, see: Max Martin, "Schmuck und Tracht des frühen Mittelalters," in *Frühe Baiern im Straubinger Land. Gaudemuseum Straubing*, Max Martin and Johannes Prammer, eds. (Straubing: Druckerei Bertsch, 1995), 40–58.

61. Heiko Steuer, *Frühgeschichtliche Sozialstrukturen in Mittelauropa: Eine Analyse der Auswertungsmethoden des archäologischen Quellenmaterials,* Abhandlungen der Akademie der Wissenschaften in Göttingen, philologisch-historische Klasse, 3d ser., vol. 128 (Göttingen:Vandenhoeck & Ruprecht, 1982), 52–3; 72–4; Edward James, "Burial and Status in the Early Medieval West," *Transactions of the Royal Historical Society* 5th ser., 39 (1989): 31–7; Heinrich Härke, "'Warrior Graves'? The Background of the Anglo-Saxon Weapon Burial Rite," *Past and Present* 126 (1990): 42–3.

62. Jürgen Hannig, "*Ars donandi:* Zur Ökonomie des Schenkens im früheren Mittelalter," in *Armut, Liebe, Ehre: Studien zur historischen Kulturforschung,* Richard van Dülmen, ed. (Frankfurt: Fischer Taschenbuch Verlag, 1988), 11–15; Hardt, "Royal Treasures," 277–80.

63. Cécile Barraud, Daniel de Coppet, André Iteanu, and Raymond Jamous, *Of Relations and the Dead: Four Societies Viewed from the Angle of their Exchanges,* Stephen J. Suffern, trans. (Oxford: Berg Publishers, 1994), 61–5; 103–5.

64. See, for instance, *Lex Burgundionum* 51, 3: "Ornamenta quoque vestimenta matronalia ad filias absque ullo fratris fratrumque consortio pertinebunt; quod quidem de his ornamentorum vestimentorumque speciebus circa filias ex lege servabitur, quarum mater intestata decesserit. Nam si quid de propriis ornamentis vestibusque decreverit, nulla in posterum actione causabitur." Ludwig Rudolf de Salis, ed., *Leges Burgundionum, MGH: Leges* 2, 1 (Hanover: Impensis bibliopolii Hahniani, 1892), 84. Other examples include the *Leges Burgundionum* 86, 1–2; *Lex Thuringorum* 31–3; *Lex Francorum Chamavorum* 42. S. Rietschel, "Heergewäte und Gerade," in *Reallexikon der Germanischen Altertumskunde* 2, Johannes Hoops, ed. (Strasbourg:Verlag von Karl J.Trübner, 1913), 467.

65. Karl August Eckhardt, ed., *Pactus legis Salicae* C27, 34–5, *MGH: Leges* 4, 1, rev. ed. (Hanover: Impensis bibliopolii Hahniani, 1962), 109.

66. Karl Friedrich von Richthofen, ed., *Lex Thuringorum* 28, *MGH: Leges* 5 (Hanover: Impensis bibliopolii Hahniani, 1875–1889), 131.

67. Gabriele von Oldberg, "Aspekte der rechtlich-sozialen Stellung der Frauen in den frühmittelalterlichen Leges," in *Frauen in Spätantike und Frühmittelalter: Lebensbedingungen-Lebensnormen-Lebensformen,* Werner Affeldt, ed. (Sigmaringen: Jan Thorbecke Verlag, 1990), 223–7.

68. Franti_ek Graus, *Volk, Herrscher und Heiliger im Reich der Merowinger: Studien zur Hagiographie der Merowingerzeit* (Prague:Tschechoslawakische Akademie der Wissenschaften, 1965), 162–165; 203; Alexander Bergengruen, *Adel und Grundherrschaft im Merowingerreich,* Vierteljahrschrift für Sozial- und Wirtschaftsgeschichte Beihefte 41 (Wiesbaden: Franz SteinerVerlag, 1958), 171–83.

69. Joachim Werner, *Das alamannische Fürstengrab von Wittislingen,* Münchner Beiträge zur Vor- und Frühgeschichte 2 (Munich: C. H. Beck'sche Verlagsbuchhandlung, 1950), 1–66.

70. But see: Max Martin, "Die Gräberfelder von Straubing-Bajuwarenstraße und Straßkirchen: Zwei erstrangige Quellen zur Geschichte der frühen Baiern im Straubinger Land," in *Frühe Baiern im Straubinger Land,* 18–27.

71. Siegfried Rietschel, "Der 'Todtenteil' in Germanischen Rechten," *Zeitschrift der Savigny-Stiftung für Rechtsgeschichte,* germanistische Abteilung 32 (1911): 297–312. On the grounds for these interpretations, see: Bonnie Effros, *Merovingian Mortuary Archaeology and the Making of the Early Middle Ages* (Berkeley: University of California Press, in press).

72. I owe great thanks to Richard Gerberding, Guy Halsall, Jörg Jarnut, Isabel Moreira, Peter Potter, and Walter Pohl for their insights into and helpful comments on this chapter. The content of this piece was derived from an unpublished part of my dissertation (UCLA, 1994), and extensive revisions were made possible by an Assigned Time for Research grant and a Summer Research Fellowship (2001) from Southern Illinois University at Edwardsville.

CHAPTER 2

FROM SELF-SUFFICIENCY TO COMMERCE:
STRUCTURAL AND ARTIFACTUAL
EVIDENCE FOR TEXTILE MANUFACTURE
IN EASTERN ENGLAND IN
THE PRE-CONQUEST PERIOD

Nina Crummy

Introduction

The basic needs of *homo sapiens* are food, water, and shelter, and in some climates shelter means not just a dry cave or hut, but portable shelter—clothes. It is no surprise, therefore, that artifacts used during the various clothmaking processes are found on archaeological sites from the Neolithic period onward. For the most part, but varying from period to period, these artifacts take the form of loom weights, spindles and their whorls, weaving tablets, tools such as weft-beaters and swords, shears and wool-combs. The recovery of the textiles themselves is much less common, depending as it does on particular conditions of deposition.

By concentrating on three particular sites, West Stow in Suffolk, Goltho in Lincolnshire, and the Coppergate excavations in York, Yorkshire, this chapter seeks to describe the change in textile production in England from self-sufficiency in the Pagan Saxon period to commercial production by the eleventh century.[1] Before turning to this, however, I briefly describe both the burial conditions necessary for textiles to survive, and also some of the artifacts used for textile manufacture in the Anglo-Saxon period, as I will refer to these in the main part of the chapter.

Preservation

The recovery of textiles is rare in Britain, as the special conditions needed for the preservation of buried organic material are not often encountered. In dry climates (hot dry or cold dry) ancient textiles have a far greater chance of survival. The tomb of Tutankhamun in Egypt, for example, produced many linen items, such as tunics, gloves, and rolls of cloth, as well as even more delicate organic items such as wreaths and garlands of leaves, flowers, berries and other fruit, while at the other extreme of temperature, in the high Alps, clothes made of leather and fur and together with a woven grass cape were recovered from the frozen body of Ötzi, the Neolithic Iceman. Neither of these special dry conditions pertains in Britain, but the island's climate provides another alternative for preservation, abundant water. Textiles buried in waterlogged soil with no free oxygen (anaerobic conditions) will survive. Well-known Scandinavian examples of textiles preserved in this way include the clothing from bog-bodies. The Early Iron Age woman found in Huldre Fen, Denmark, had a woven scarf and a check skirt, patterned in dark and light brown wool (probably not the original colors), while Bocksten man from Sweden wore trousers, a tunic, a cape, and a distinctive hood with a long point at the back that allowed the date of his murder and clandestine burial to be placed around 1360 A.D.[2]

Unfortunately for textile researchers, many bog-bodies were sacrifices and buried naked, as was the case with Iron Age Lindow man, a.k.a. "Pete Marsh," from Lindow Moss peat bog in Cheshire. He had been effectively killed four times: his skull had been smashed, he had been garrotted, his throat had been cut, and finally, probably already dead, he was symbolically drowned.[3] A lost opportunity to examine a fully-clothed Romano-Briton occurred when the body of a man was found by peat-diggers on Grewelthorpe Moor in 1850. He was no sacrifice, but must have drowned in the bog while trying to cross the moor. A contemporary report of the discovery described him as wearing a green cloak, a tunic at least partly scarlet, and yellow stockings. Unfortunately, by the time the local policeman reached the scene, he was only able to recover the stockings and shoes. Only two scraps of the fabric now survive, both the pale brown to which buried woolens usually revert. One is an insole cut to fit in the only surviving shoe, the other is probably from one of the stockings.[4] The waterlogged archaeological deposits of London and York have yielded many fragments of textiles, mainly dating from the ninth century onward.[5] The pieces from the Coppergate site, York, are perhaps the most remarkable, as they include a sock made by *nålebinding* (a form of knitting using only one needle, producing a thick elastic fabric that will not unravel if damaged), a silk headdress, and a silk reliquary pouch.[6]

There are other ways in which cloth can be preserved. Slow burning can carbonize textile, which happened to diamond twill and plain weave fabrics on a bed burnt in Roman Colchester during the Boudican revolt of 60/1 A.D.[7] Small scraps of fabric can also survive when buried in close association with metal objects. If a person is buried in a cloak fastened by an iron brooch, most of the garment will rot away, but where the corroding brooch touches the textile it replaces the fiber with ferrous oxide, allowing a mineralized piece of fabric to survive. At Edix Hill, Barrington, Cambridgeshire, the site of a Migration period cemetery, over 270 "replaced" textile fragments have survived from nineteenth and twentieth-century excavations on the site.[8] Similarly, at Morning Thorpe cemetery, Norfolk, nearly 300 textile fragments were recovered.[9] Both sites provided evidence of tablet-woven cords and braids, tabbys, and twills.

Textile can also be preserved is as an impression in plaster. A few wealthy Romano-Britons of the late third and fourth centuries were buried in coffins into which lime plaster or gypsum had been poured. Where the bones pressed firmly against the plaster, e.g., at the shoulder blades, impressions can be made of the weave of the textile garments in which the dead person was buried.[10]

The fiber used in the tiny fragments preserved in these three ways can rarely be identified, but the weave of the fabric, the direction in which the thread was spun can usually be seen. They may even provide evidence of trade, as a tiny fragment of a plain weave unspun silk from a late Roman burial at Butt Road, Colchester, probably came from China.[11]

Thread can also be metal. Gold thread has a good chance of survival, though it is usually fine and brittle. Gold tissue was found in the Lexden Tumulus at Colchester, dated to ca. 10 B.C. and possibly the burial mound of Addedomaros, leader of the Iron Age tribe of the Trinovantes.[12] Tiny fragments of gold thread have also been found in late Roman burials of high-status young women at Winchester and London,[13] in several well-furnished early Anglo-Saxon graves in England and contemporary burials on the continent,[14] on the vestments of St Cuthbert,[15] and, in a more mundane eleventh-century setting, on the urban Coppergate site at York.[16]

Anglo-Saxon Artifacts
from Textile Manufacture

In contrast to the rarity of the end product itself in the archaeological record, many artifacts used during the various processes of textile manufacture, in particular spinning and weaving, have been found on Anglo-Saxon excavations in Britain. Clay items such as loom weights and spindlewhorls are rarely subject to decay, and stone, a popular material for spindlewhorls, is also usually stable, though the surface of the softer rocks may discolor or weather. Iron tools such as wool-combs corrode badly when buried, but X-radiography and conservation have helped identify and preserve a substantial number. Bone objects such as weaving combs and pinbeaters may be affected by acidic soil, but they are mostly well preserved. Spindles and looms were made of wood, which, like textile, is only preserved in very dry or wet sealed conditions, or when converted to charcoal.

Simply a ring of clay twisted into a circle, loom weights are common on occupation sites of the fifth to seventh centuries, and were not necessarily fired before use. The unfired loom weights from Willington, Derbyshire, were presumed to have been used in "green."[17] Unfired weights were found at Pennylands, Buckinghamshire,[18] and Mucking, Essex.[19] The assemblage from West Stow, Suffolk, included fired, partially fired, and unfired loom weights,[20] and it may be that weights were made some time before firing and stored until required, though this does not explain partly-fired weights. Loom weights changed little over the Anglo-Saxon period, though by the tenth century they were more bun-shaped than annular.[21] With the replacement of the warp-weighted loom by the more advanced horizontal loom in the eleventh century, the loom weight disappears from the archaeological record.

Figure 2.1a is a double-ended bone pinbeater and Figure 2.1b is a single-ended example. Both tools were probably used to separate the warp threads between throws on a warp-weighted loom, though there single-ended ones may also have been used on the vertical two-beam loom.[22] The double-ended form originated on the continent during the early Roman period, with the single-ended examples not appearing until about the ninth century. They are always highly polished to prevent them from catching on the threads.

Figure 2.1c is a pig fibula with the proximal end removed and the distal end modified to form a simple triangular head. The precise function of these objects has been the subject of much discussion in the archaeological literature, partly due to the wide variety of forms discovered. The heads may be triangular or rounded, and Late Saxon examples can be ornamentally carved. They may be pierced or not. They have been variously interpreted as dress pins,[23] awls,[24] and needles,[25] but Ian Riddler has recently noted that they are often found in contextual association with loom weights and pinbeaters, and he suggests that they should also be viewed as tools associated with weaving on a warp-weighted loom.[26] However, the shanks and heads of very few of these objects have the high degree of polish shown by the pinbeaters, suggesting that only the tip and the area immediately above came into contact with the thread.

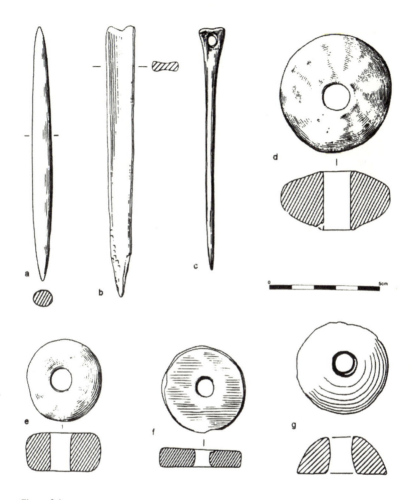

Figure 2.1
Tools associated with spinning and weaving. Anglo-Saxon England. Drawings by Kate Crummy.

   a) Double-ended Pinbeater. Bone. Early Roman period.
   b) Single-ended Pinbeater. Bone. Circa ninth Century.
   c) Tool or ornament. Bone: pig fibula.
   d) Spindle whorl. Fired clay. Pagan Saxon period.
   e) Spindle whorl. Stone (fine-grained mudstone or hard "chalk").
   f) Spindle whorl. Pierced recycled shard of hard-fired late Roman pottery.
   g) Spindle whorl. Bone. Head of cattle femora.

Spindlewhorls could be made of fired clay (Figure 2.1d), stone, often fine-grained mudstones or hard "chalk" (Figure 2.1e), pierced recycled shards of hard-fired late Roman pottery (Figure 2.1f), and the heads of cattle femora (Figure 2.1g). The spindles would have been wooden and few have survived, though four fragments have been recovered from York.[27] The range of lengths and maximum diameters shown by the York spindles demonstrates that both fine and thicker yards were spun at Coppergate. Where no spindle survives, the diameters of the hole in a whorl can show the spindle diameter and thus the yarn grade. At Pennylands the hole diameters were small, restricted to between eight to ten millimeters, while at West Stow they ranged from five to 16 mm and at Mucking of from six to 19 mm.[28]

The wider variety of objects associated with textile manufacture recovered in the Late Saxon period is much greater than from the Pagan Saxon period, and to some extent this must be due to the fact most of the earlier sites "metal-poor." In a largely subsistence and exchange economy, metal ores were hard to come by, especially in some areas, so broken and worn-out objects had a high "scrap" value and were usually melted down. The rise of towns in the Middle Saxon period meant a move away from subsistence to a demand for manufactured goods, resulting in more waste, thereby enriching the archaeological record. Thus, at West Stow village, metal weaving equipment is limited to a weaving batten,[29] while at Anglo-Scandinavian Coppergate, in urban York, there are wool-combs, a weaving batten, heckles, weft-beaters, tenterhooks, shears, and a shearboard hook.[30] Over the last 20 years the number of individual wool-comb or heckle teeth recorded from archaeological sites has risen markedly, thanks to X-radiography. Covered in corrosion they simply resemble structural nails, but X-rays reveal them to be long and thin, with a round or rectangular rather than square section.[31] The waterlogged soil at Coppergate also aided the preservation of vegetable matter, producing woad, madder, greenweed, and weld, all used for dyeing, and also a few fragments of the flowers and fruits of the teasel, *Dipsacus sativus,* used to raise a nap on cloth.[32]

Self-Sufficiency at West Stow

No towns, as such, existed in Late Iron Age Britain at the time of Roman invasion of 43 A.D., the nearest equivalent being the loose agglomerations known as *oppida.* A characteristic of the finds assemblages from sites at this period is the quantity of loom weights recovered (Iron Age loom weights were not annular, like Anglo-Saxon examples, but triangular). Their high number and ubiquity suggests that textile production was carried out by all communities, and probably all families, regardless of status: the Iron Age Britons were self-sufficient.

It is usually assumed that the triangular Iron Age loom weights were replaced soon after the Roman invasion by Roman pyramidal loom weights,[33] but in fact loom weights are rarely found on Romano-British sites, which suggests that weaving ceased to be an everyday occupation of most households, and became instead factory or workshop-based. This pattern is repeated in the eleventh century, with loom weights disappearing in the eleventh century along with the warp-weighted loom, to be replaced by the faster, more "industrial," horizontal loom.

When the Roman legions were withdrawn from the province in the early fifth century and the number of settlers coming across the North Sea increased, at least in eastern Britain, the civil administration probably collapsed fairly rapidly so that towns and the industries that catered to their needs no longer functioned. Britain returned again to a largely agrarian way of life centered on small communities, be they native (Romano-) British, Anglo-Saxon, or a mixture of both. The archaeological record from excavations carried out on Migration period sites tells of self-sufficient farmers providing for their own needs, building their houses from naturally-available materials, growing crops, tending beasts, and making a limited range of domestic objects and tools.

The Anglo-Saxon village of West Stow in Suffolk lies on the light sand and gravels of the Breckland, on a small knoll close to the River Lark. The site was intensively excavated between 1957 and 1972, and soil sample and animal bone analyses have allowed a detailed picture to be built up of the village and surrounding area from the early fifth to the mid-seventh centuries. Most of the valley was open arable farmland with plentiful pasture and stands of willow, hazel, alder, ash, and hawthorn, with oak scrub or forest nearby to provide building material. Hens, ducks, and geese foraged for food among the buildings of the village, or around the pigsty. Spelt, barley, rye, and bread or club wheat grew in the surrounding fields, and sheep and cattle grazed on the valley slopes. There were red and roe deer and wild fowl to be hunted,[34] and the River Lark provided fish.

There was a limited amount of iron-smithing within the community; bone combs and tools were made, some to be used in spinning or weaving, and many clay loom weights also point to weaving.[35] No textiles were recovered from the acidic soil, though a small number of metal objects had mineralized fragments of Z- and S-spun thread attached, tiny pieces of cloth identified as plain weave, half-basket weave, and tablet twists.[36] The variety of styles of pottery suggest the community made their own pots from the local clay, though the villagers in the second half of the sixth century also bought the wares of the Illington/Lackford potter, a talented artisan who worked nearby.[37] A few fine items such as beads, brooches, and glassware point to contact with traders from farther afield, and in a bartering society these goods would probably have been bought with agricultural surplus—crops, wool, and hides.[38]

Over the 250-odd years of its life, the village consisted of about three halls surrounded by a number of huts. The halls were rectangular post-built structures, but each hut was a sunken-featured building (SFB), it had a plank floor set over a hollow flat-bottomed pit. A hall and a few huts have now been reconstructed, so that modern visitors can get some idea of what the village was like.[39]

Many of the 69 huts and two of the halls contained at least one spindle-whorl, a possible weaving batten came from one hall, and pinbeaters and modified pig fibulae were found in several huts. Twenty-two huts contained loom weights, usually fewer than would be needed for a loom, though far more in two burnt huts (SFBs 3 and 15). The evidence from these buildings, dated to the late sixth to seventh century, is interpreted as showing that the warp-weighted looms used at West Stow were free-standing, not fixed to posts sunk into the floor.[40] Seventy-three loom weights were found in SFB 3, mostly in the eastern half of the hut and resting on collapsed planks from the floor. Some lay in short rows, others in jumbled heaps.[41]

In SFB 15 there were three groups of loom weights on collapsed planks, two large (60 and 90) and one small (18). It has been argued that at least three looms were in use in the hut when it was burnt.[42] Most of the 18 weights found in the southwest corner were stacked on edge, suggesting that they dropped together from a loom during the fire. Two similar, if shorter, rows are also identifiable in the group of 60 weights in the northeast quadrant, parallel to each other but some five feet apart, with the majority of the group lying in a jumble in the corner of the hut. The largest number of weights was from the northwest quadrant, where they lay haphazardly tumbled about.[43] The implication of 90, even 60, weights is a very large loom, perhaps too long for a hand-shuttle to be passed across.

Do rows of weights necessarily imply the position of a loom? At Grimstone End, Pakenham, Suffolk, about eight miles east of West Stow, two sets of loom weights, Series "A" and "B" were found. The 62 weights comprising Series "B" lay close together in two rows about eight feet long (2.5 meters), eight or nine inches apart, and converging at one end.[44] The excavators rejected the idea that Series "B" represented a collapsed loom, arguing that eight feet was too far to throw a hand-shuttle, but suggesting instead that the weights were in the position in which they had been fired, an argument backed up by a nearby heap of wood and charcoal and by traces of wood beneath the weights.[45] Despite these very reasonable caveats, the group has been cited as representing a loom.[46] Another batch of 100 or so loom weights from the Middle Saxon settlement of Aldwych, London, was also presumed to be the site of a loom, despite the fact that again this number implies a very large loom, too long for throwing a hand-shuttle.[47]

At Old Erringham, West Sussex, lines of weights were not considered sufficient proof alone for proposing the site of a loom,[48] and at Upton, Northamptonshire, rows of weights with traces of wood inside the holes were considered to be evidence that weights were stored on a stick or pole when not in use.[49]

A positively identified loom found in an eleventh-century hut at Back Street, Winchester, had used only 22 or 23 loom weights.[50] The group lay on edge in a row, 1.7 meters long, between a single posthole and a double posthole.[51] Given the association with the postholes, this loom was fixed, not free-standing. Like those of West Stow and Grimstone End the weights are poorly fired. More were found in a heap near one end of the row and others lay scattered over the floor of the hut.[52] The group is comparable to the 18 from SFB 15 at West Stow, and the notion that spare weights were stored close at hand allows the large groups at West Stow to representing both a loom (or looms) and spares. At Bourton-on-the-Water, Gloucestershire, an early Anglo-Saxon hut appears to have held an upright loom fixed between two posts set in the ground seven feet, nine inches apart (center to center), with a stone seat for the weaver set on a clay ramp in front of it. About 70 loom weight fragments came from this hut.[53]

Another possible interpretation of the rows of Series "B" at Grimstone End is that they were strung on sticks to dry before firing.[54] The weights from SFBs 3 and 15 at West Stow were unfired, with burning only on the exposed upper surface. The excavator suggested this showed that they were used "green,"[55] but there must be a strong possibility that the West Stow huts were the site of loomweight manufacture or storage.

Whatever the interpretation of the groups of weights at West Stow, it is clear that the community spun wool and wove cloth. The high number and wide distribution of loom weights and spindlewhorls from West Stow, together with the pinbeaters and weaving batten, was initially taken to suggest that weaving was a major economic support of the village, in that there was a large surplus for trade.[56] This was a rather "enthusiastic" interpretation made by first impressions of the finds assemblage. More detailed post-excavation analysis tempered this view to that expressed in the final published report, where it was suggested that meat and leather production had equal importance with weaving.[57]

However, the animal bone report for West Stow presents a different point of view. Looking at the kill patterns of sheep from a site provides information about the use of the flock. More than one third of the village's sheep were killed between the ages of six and twelve months, half were dead before they were one year old, and two-thirds before they were two.

A second peak of killing occurred when the sheep were between the ages of four and six years. Only six percent of the sheep population lived to more than six years of age. Sheep kept for wool production live longer, so this kill pattern is closest to one for a flock bred for meat and milk and inappropriate for wool, though the kill-off in the four-to-six year olds suggests some sheep were kept alive for longer both for breeding and so that their wool could be collected for domestic use.[58] So West Stow appears to have produced enough cloth for its own needs, and no more.

The frequency of retrieval, whether in a hut or not, of the occasional loom weight, and/or spindlewhorl, and/or pinbeater as seen at West Stow is repeated throughout the Anglo-Saxon period in England across many sites, pointing to the frequency of low level "self-sufficient" spinning and weaving, for example: Colchester,[59] Gamlingay,[60] Godmanchester,[61] Ipswich,[62] Keston,[63] Lincoln,[64] London,[65] Maxey,[66] Mucking,[67] Northampton,[68] Oxford,[69] Pennylands,[70] St. Neots,[71] Seacourt,[72] Shakenoak,[73] Southampton,[74] Thetford,[75] Upton,[76] and Winchester.[77]

Surplus Production at Goltho

Some two centuries after West Stow was abandoned, there is good evidence for the manufacture of surplus cloth at the manor of Goltho, about nine miles northeast of Lincoln, where a succession of buildings have been identified as weaving sheds. Goltho, a fortified enclosure constructed ca. 850, lies on a bed of Boulder Clay, with the enclosure (later developed into a motte-and-bailey castle) sitting slightly higher than the village on a thin layer of sand. Poor drainage gives the heavy clay soils a tendency to become waterlogged and compacted. A spell of rain, not necessarily prolonged, coupled with slow evaporation can make cultivation of the soil difficult, even impossible. If it is to crop well, such soil is best given long periods of rest under grass, which can then be used for grazing sheep.[78]

The Late Saxon manor was established on the site of a Middle Saxon homestead, and remained in the same position until ca. 1150, when the site was abandoned. It initially consisted of a strong rampart and ditch enclosing a subsquare area about 48 meters by 48 meters. Excavation of the interior revealed the remains of a range of substantial timber buildings constructed around three sides of a courtyard. The focal point of the enclosure was a bow-sided hall, comparable in size to the eight-posthole hall, possibly built to accommodate meetings of the *witan,* at the Royal Palace site at Cheddar, Somerset.[79] At right angles to the hall lay the kitchen, and tucked into the southeast corner was the bower. Opposite the hall lay a long straight-sided building, 50 to 60 feet (15.6–18 meters) long by 15 feet (4.5 meters) wide, identified as a weaving shed.[80] Over the years all these buildings were subject to alterations and rebuilding, with the periods of interest for textile manufacture dated by the excavator to ca. 850–ca. 950 (Period 3), and ca. 950–ca. 1000 (Period 4).

Thirty-five artifacts associated with spinning and weaving were found at Goltho, 29 of them in specific late ninth- and tenth-century contexts. Compared to the individual halls and huts at West Stow, this is quite a high finds recovery rate, despite very few refuse pits being found and the disturbance of early levels by clearance for later buildings. Thirteen spindlewhorls were found, mostly from the hall and bower, demonstrating that spinning was not confined to specialist workers and was carried out by the female population in general, up to, and probably including, the lady of the manor. In contrast, a shearboard hook, used to fix cloth to a cropping-board for finishing, six pinbeaters and fourteen wool-comb teeth were found concentrated in the weaving shed and the part of the courtyard adjacent to it.[81] Clearly combing, weaving, and finishing were carried on here.

The pinbeaters indicate the use of warp-weighted vertical looms, but the absence of loom weights suggests that the looms had a horizontal warp beam at the base. Such looms can be set into the floor, leaving long trenches to define their position. Slots and postholes cut into the floors of the shed were found, but they could equally well have been from internal partition walls as much as looms. A cesspit lay about three feet from the northeastern corner of the shed, and shards from cooking pots and bowls were recovered from its construction and occupation levels, suggesting that the shed also provided domestic accommodation for the weavers. This is supported by the continental *Leges Alamannorum,* which stipulated a fine of six shillings for anyone violating 'a maiden from the weaving-shed.[82]

The animal bone sample at Goltho (though flawed) reveals an interesting change in the numbers of sheep/goat killed from ca. 850–ca. 1000 compared to later periods. The late ninth to tenth century shows a low use of sheep/goat for meat, 16.2 percent compared to 52.5 percent for cattle, while in the eleventh century, when the weaving shed was replaced with domestic buildings, the use of sheep/goat rises slightly to 16.8 percent compared to 27.1 percent for cattle, and in the late eleventh to early twelfth century sheep/goat use rises to 23.8 percent compared to 10.6 percent for cattle. Unfortunately no age data has been made available, but the low kill rate over the period when weaving was practiced suggests flocks kept for wool not meat.[83]

Commerce at Coppergate

At the same period that the Goltho weaving sheds were in use, some 50 or 60 miles to the north the Anglo-Scandinavian town of York was flourishing.

Both a royal and a religious site, York owed its survival and success to trade, with good communications into a rich hinterland and across the North Sea to Europe. In 866 it was taken by the Vikings, who made it, as Jorvik, the capital of an independent kingdom. It was subsumed into the new creation of England in 954.

The Coppergate site at York lies in the north-eastern part of a tongue of land between the rivers Ouse and Foss, bounded on the east by the banks of the Foss, on the north by the street known now as Piccadilly, and on the west by the street of Coppergate, a street that led in the medieval period to the only bridge across the Ouse.[84] The site was vacant land in the Anglian period, and from the mid-ninth century to ca. 930/5 (Periods 3 and 4a) it appears to have been used for rubbish disposal, with some post- and stakeholes suggesting boundary alignments based on possible new buildings fronting Coppergate, which was laid out in the early tenth century.[85] Organic material was preserved in the rubbish pits, and the finds included dyed raw wool, spindlewhorls, loom weights, shears, scraps of woolen textiles, linens, woolen cords and iron needles. This complete range of items demonstrates that textiles were manufactured and processed from fleece right through to completed garments, if not on the site then immediately adjacent to it.[86]

In Period 4B, dated ca. 930/5 to ca. 975, four tenements were clearly distinguishable, with post-and-wattle buildings fronting Coppergate. They were vulnerable to fire as well as decay, and were repaired or replaced frequently. Metal working was the main activity, with objects of iron, copper-alloy, lead-alloy, gold, and silver being manufactured. While metal working may have been predominant, the high number of wool-comb and flax heckle teeth from the site also point to fiber-processing on the site, with their distribution confirming that wool-combing was an indoor, heckling an outdoor, activity. Spinning and weaving were also carried out, and there is an abundance of evidence for dyeing, and for sewing or repairing garments.[87]

Organic-rich occupation deposits accumulated rapidly in this period, both in and around the buildings and in pits.[88] Many pieces of raw wool were found, as well as a bundle of horse-tail hair, lengths of woolen yarn and cord, silk yarn, numerous scraps of tabbys and twills, both woolen and linen, a fragment of a silk and linen tablet-woven braid, and many fragments of silk tabbys, with one possible piece of silk twill.[89] An unused offcut of silk points to the importation of bolts of the cloth to be made up into garments on the site.[90] At Coppergate a similar quantity and range of items came from Periods 5A and 5B (ca. 975 to the early/mid-eleventh century), when semi-basements were inserted into the tenements.[91] Several fragments, unidentifiable in the earlier periods, proved to be of white linen.[92]

The majority of the Coppergate textiles appear to be unwanted pieces of everyday items, clothes, soft furnishings, and sacks, along with scraps such as yarn ends left over from manufacturing cloth.[93] Most are badly soil-stained, but analysis revealed the use of dyes made from madder (brick-red color), woad (blue), kermes (bright red; an imported dye produced from insect bodies), and lichen purple.[94] There is also evidence of madder dyeing in the Late Saxon period from Thetford[95] and London, where 15 dyepots were associated with pottery of the pre-Conquest period,[96] and cloth fragments showed the use of madder, woad, and lichen purple.[97] Wool was usually dyed in the fleece or the skein, rather than as woven cloth, permitting stripes or checks to be produced.[98]

At Coppergate artifacts associated with spinning and weaving included large numbers of spindles, spindlewhorls, pinbeaters, a weaving tablet, glass linen smoothers and some loom weights. The catalogue of the iron objects includes a wool-comb fragment, a weft-beater fragment, nearly 200 individual wool-comb or flax heckle teeth, 12 shears or shear fragments, most from Periods 4B and 5B contemporary with the numerous textiles, and a possible shearboard hook.[99] Walton Rogers has stressed the even chronological and chorological spread of textile equipment at Coppergate, where it occurs across all periods and across all four tenements, but has also noted that there appears to be more clothmaking activity on the site than might be considered usual for self-sufficiency.[100] Indeed, the occupants of Coppergate, who would have had to buy in their raw materials, could not be called "self-sufficient," though all the textile crafts from fiber preparation, spinning, dyeing, weaving, cutting, and sewing appear to have been practiced on the site at one time or another, and in York's growing population there must have been a ready market for any surplus.

The absence of loom weights from the warp-weighted loom after Period 5B at Coppergate is matched by the appearance in Period 6 by evidence for the use of the new, faster, horizontal loom, which probably arrived in England in the eleventh century. There is no evidence for dyeing after Period 5C, though the iron spikes from wool-combs and flax heckles continue to occur in abundance and are matched by the appearance of bale pins, evidence that the raw wool was carted to the site for the first stage of processing.[101] Thus at Coppergate can be seen the beginnings of a move away from home-based production toward the specialization necessary for large-scale output and trade, the division of labor that gave rise to the medieval guilds, with carders, dyers, weavers, tailors, and so on, all operating separately.

Conclusion

The rise of commercialism in textile production can be seen as linked inextricably to that of urbanism. The farming community of West Stow had some contact with wider markets but was essentially self-sufficient. There is no evidence from the artifactual or structural record that spinning and weaving took place in purpose-built structures, rather it seems to have been carried on in the dwelling place.

The lordship of Goltho presents a different picture, for there textile production took place over a prolonged period of time in a substantial building that also housed the workers. Raw wool was produced on the demesne, and turning it into cloth was clearly an important activity in the manor's daily round. But it is impossible to tell how much the weaving-shed represents simply the providing of optimal facilities to enable all the cloth required by the inhabitants of Goltho to be supplied "in-house," and how much it was a commercial venture aiming to produce a surplus for export. Certainly, if more were made than was needed, there was easy access to a regular and demanding market at the nearby urban center of Lincoln.

At the prosperous expanding town of Viking Jorvik, the beginnings of real commercialism can be seen. The inhabitants of the Coppergate tenements were removed from the actual process of breeding and maintaining sheep and shearing them for their wool, and also from the cultivation and harvesting of flax. They purchased raw materials for conversion into finished products—and surely this passes for a definition of a factory or workshop. Wool and flax probably came from the surrounding countryside, dyestuffs probably came from the continental mainland, and silk from either the eastern Mediterranean or from China.

Notes

1. Please note that British archaeologists reserve the term "medieval" for the time from the Norman Conquest (1066 A.D.) to the late fifteenth century. The fifth and sixth centuries are usually referred to as the Pagan Saxon or Migration period, the seventh to early-ninth centuries are the Middle Saxon period, and the Late Saxon period runs from the mid-ninth century to the Norman Conquest. The term "Anglo-Saxon" covers the entire period from the fifth to the eleventh century; and "Anglo-Scandinavian" describes the distinctive culture of York from the mid-ninth century to the Conquest.

2. P. V. Glob, *The Bog People* (London: Paladin, 1971), 94–5, 110.

3. S. James and V. Rigby, *Britain and the Celtic Iron Age* (London: British Museum Press, 1997), 64.

4. R. C. Turner, M. Rhodes, and J. P. Wild, "The Roman body found on Grewelthorpe Moor in 1850: a reappraisal," *Britannia* 22 (1991): 191–201.

5. E.g. F. Pritchard, "Late Saxon textiles from the City of London," *Medieval Archaeology* 28 (1984): 46–76; E. Crowfoot, F. Pritchard, and K. Staniland, *Textiles and Clothing c. 1150–1450*, Medieval Finds from Excavations in London 4 (London: Her Majesty's Stationery Office, 1992); P. Walton, *Tex-*

tiles, Cordage and Raw Fibre from 16–22 Coppergate, The Archaeology of York 17/5 (York: Council for British Archaeology, 1989).

6. Walton, *Textiles, Cordage and Raw Fibre,* figs. 140, 151, 156.

7. P. Crummy, J. P. Wild, and J. Liversidge, "The bed" in P. Crummy, *Excavations at Lion Walk, Balkerne Lane, and Middleborough, Colchester, Essex,* Colchester Archaeological Report 3 (Colchester: Colchester Archaeological Trust, 1984), 42–7.

8. E. Crowfoot, "Textiles associated with metalwork" in T. Malim and J. Hines, *The Anglo-Saxon Cemetery at Edix Hill (Barrington A), Cambridgeshire,* CBA Research Report 112 (York: Council for British Archaeology, 1998), 235–46.

9. E Crowfoot, "Report on the textiles" in B. Green, A. Rogerson, and S. G. White, *Morning Thorpe Anglo-Saxon Cemetery, Norfolk,* East Anglian Archaeology 36, (Dereham: Norfolk Archaeological Unit, 1987), 175–88.

10. J. P. Wild 1983, "Textile fragments from the later Butt Road cemetery" in N. Crummy, *The Roman Small Finds from Excavations in Colchester 1971–9,* Colchester Archaeological Report 2 (Colchester: Colchester Archaeological Trust 1983), no. 4303.

11. Wild, "Textile fragment from the later Butt Road cemetery," no. 4300.

12. P. G. Laver, "The excavation of a tumulus at Lexden, Colchester," *Archaeologia* 76 (1927): 251, plate 62, fig. 1.

13. N. Crummy, P. Ottaway, and H. Rees, *Small Finds from the Suburbs and City Defences,* Winchester Museums Publication 6 (Winchester: Winchester City Museums, forthcoming).

14. E. Crowfoot and S. C. Hawkes, "Early Anglo-Saxon gold braids," *Medieval Archaeology* 11 (1967): 42–86.

15. G. Crowfoot, "The tablet-woven braids from the vestments of St Cuthbert at Durham," *Antiquaries Journal* 19 (1939): 57–80.

16. Walton, *Textiles, Cordage and Raw Fibre,* 314, 440, no. 1410.

17. S. M. Elsdon, "Baked clay objects: Anglo-Saxon," in H. Wheeler, "Excavations at Willington, Derbyshire, 1970–72," *Derbyshire Archaeological Journal* 99 (1979): 210.

18. R. J. Williams, *Pennyland and Hartigans,* Buckinghamshire Archaeological Society Monograph 4 (Aylesbury: Buckinghamshire Archaeological Society, 1993), 123.

19. H. Hamerow, *Excavations at Mucking 2: the Anglo-Saxon settlement,* English Heritage Archaeological Report 21 (London: English Heritage, 1993), 68.

20. S. West, *West Stow, the Anglo-Saxon Village,* East Anglian Archaeology 24, (Ipswich: Suffolk County Council Planning Department, 1985), 138.

21. P. Walton Rogers, *Textile Production at 16–22 Coppergate,* The Archaeology of York 17/11 (York: Council for British Archaeology, 1997), fig. 813.

22. Walton Rogers, *Textile Production at 16–22 Coppergate,* 1759–60.

23. A. MacGregor, *Anglo-Scandinavian Finds from Lloyds Bank, Pavement, and other sites,* The Archaeology of York 17/3 (London: Council for British Archaeology, 1982), 91–2.

24. S. West, *West Stow, the Anglo-Saxon Village,* fig. 30, 14.

25. J. E. Mann, *Early Medieval Finds from Flaxengate,* The Archaeology of Lincoln 14/1 (London: Council for British Archaeology, 1982), 25–6.

26. I. Riddler, "Saxon worked bone objects" in Williams, *Pennyland and Hartigans,* 114; I. Riddler, monograph on the bone objects from excavations in Ipswich, for Suffolk County Council, forthcoming.

27. Walton Rogers, *Textile Production at 16–22 Coppergate,* fig. 804.

28. Williams, *Pennyland and Hartigans,* 119–21; West, *West Stow, the Anglo-Saxon Village,* fig.30 *passim;* Hamerow, *Excavations at Mucking 2,* 64–5.

29. West, *West Stow, the Anglo-Saxon Village,* fig.21A.

30. P. Ottaway. *Anglo-Scandinavian Ironwork from Coppergate,* The Archaeology of York 17/6 (London: Council for British Archaeology, 1992), 538–51; Walton Rogers, *Textile Production at 16–22 Coppergate,* 1727, *passim.*

31. F. Pritchard, "Small Finds" in A. Vince (editor), *Finds and Environmental Evidence,* Aspects of Saxo-Norman London 2, London and Middlesex Archaeological Society Special Paper 12 (London: London and Middlesex Archaeological Society, 1991), 135, fig.3.15; Ottaway, *Anglo-Scandinavian Ironwork from Coppergate,* 538–41; Walton Rogers, *Textile Production at 16–22 Coppergate,* 1727–31.

32. A. R. Hall, "Teasels" in Walton Rogers, *Textile Production at 16–22 Coppergate.*

33. J. P. Wild, *Textile Manufacture in the Northern Roman Provinces* (Cambridge: Cambridge University Press, 1970), 63; G. Lambrick and M. Robinson, *Iron Age and Roman riverside settlements at Farmoor, Oxfordshire,* Oxfordshire Archaeological Unit Report 2, Council for British Archaeology Research Report 32 (London: Council for British Archaeology, 1979), 57.

34. West, *West Stow, the Anglo-Saxon Village,* 1, 169; P. Crabtree, "The faunal remains" in West, *West Stow, the Anglo-Saxon Village,* 86.

35. West, *West Stow, the Anglo-Saxon Village,* 138–40, 169.

36. E. Crowfoot, "The textiles" in West, *West Stow, the Anglo-Saxon Village,* 69–70.

37. J. N. L. Myres, *Anglo-Saxon pottery and the settlement of England* (Oxford: Clarendon Press, 1969), 132–6.

38. West, *West Stow, the Anglo-Saxon Village,* 169–70.

39. West, *West Stow, the Anglo-Saxon Village,* 168–9.

40. West, *West Stow, the Anglo-Saxon Village,* 138–40, 150, 181–4.

41. West, *West Stow, the Anglo-Saxon Village,* 16, fig.35.

42. West, *West Stow, the Anglo-Saxon Village,* 138, fig.71.

43. West, *West Stow, the Anglo-Saxon Village,* 23.

44. B. J. W. Brown, G. M. Knocker, N. Smedley, and S. E. West, "Excavations at Grimstone End, Pakenham," *Proceedings of the Suffolk Institute of Archaeology* 26, pt. 3 (1954): 198–9, fig.23, pl. 24.

45. Brown et al, "Excavations at Grimstone End, Pakenham."

46. J. W. Hedges, "The Loom-weights" in J. Collis, *Winchester Excavations II: 1949–60. Excavations in the Suburbs and the Western part of the Town,* (Win-

chester: Winchester City Museums, 1978), table 3; West, *West Stow, the Anglo-Saxon Village,* 138.

47. Pritchard, "Small Finds," 167.
48. E. W. Holden, "Excavations at Old Erringham, Shoreham, West Sussex," *Sussex Archaeological Collections* 114 (1976): 309, pl. 4.
49. D. A. Jackson, D. W. Harding, and J. N. L. Myres, "The Iron Age and Anglo-Saxon site at Upton, Northamptonshire," *Antiquaries Journal* 49 (1969): 210.
50. Hedges, "The loom-weights," 33–9.
51. Hedges, "The loom-weights," pl. 1a, fig.12, F10, F7, F8.
52. Hedges, "The loom-weights," 33, fig.12, F9, F5.
53. G. C. Dunning, "Bronze Age settlements and a Saxon hut near Bourton-on-the-Water, Gloucestershire," *Antiquaries Journal* 12 (1932): 284–7, 290.
54. Brown et al, "Excavations at Grimstone End, Pakenham," 190, 198.
55. West, *West Stow, the Anglo-Saxon Village,* 138.
56. S. West, "The Anglo-Saxon village of West Stow: an interim report of the excavations," *Medieval Archaeology* 13 (1969): 1–20.
57. West, *West Stow, the Anglo-Saxon Village,* 138.
58. Crabtree, "The faunal remains," 93.
59. P. Crummy, *Aspects of Anglo-Saxon and Norman Colchester,* Colchester Archaeological Report 1, Council for British Archaeology Research Report 39 (Colchester: Colchester Archaeological Trust, and London: the Council for British Archaeology, 1981), 4.
60. N. Crummy, *The Small Finds from Gamlingay,* for Hertfordshire Archaeological Trust (HAT 211/257) forthcoming (a).
61. N. Crummy, *The Small Finds from Godmanchester,* for Hertfordshire Archaeological Trust (HAT 339) forthcoming (b).
62. Riddler, monograph on the bone objects from excavations in Ipswich, forthcoming.
63. B. Philp, *Excavations in West Kent, 1963–70* (Dover: Kent Archaeological Rescue Unit, 1973), 156–63.
64. Mann, *Early Medieval Finds from Flaxengate,* 22–5.
65. Pritchard, "Late Saxon textiles from the City of London," 63–5; Pritchard, "Small Finds," 165, 167–8, 203–5.
66. P. V. Addyman, "A Dark Age settlement at Maxey, Northants," *Medieval Archaeology* 8 (1964): 58, fig.12, 15, fig.16, 21, 22.
67. Hamerow, *Excavations at Mucking 2,* 64–8.
68. J. H. Williams, *St Peter's Street, Northampton, excavations 1973–6* (Northampton: Northampton Development Corporation, 1979), fig.21, 12–13.
69. F. Radcliffe, "Excavations at Long Lane, Oxford," *Oxoniensia* 26–7 (1963): fig.15.
70. Various specialist reports in Williams, *Pennyland and Hartigans.*
71. P. V. Addyman, "Late Saxon settlement in the St Neots area, 3: the village or township at St Neots," *Proceedings of the Cambridge Antiquarian Society* 64 (1973): 90–1.

72. M. Biddle, "The deserted medieval village of Seacourt, Berkshire," *Oxoniensia* 26–7 (1963): fig.32.

73. A. C. C. Brodribb, A. R. Hands, and D. R. Walker, *Excavations at Shakenoak Farm, near Wilcote, Oxfordshire* 3 (Oxford: privately printed, 1972), 48, figs.62–4.

74. P. V. Addyman and D. H. Hill, "Saxon Southampton: a review of the evidence; Part II: Industry, trade and everyday life," *Proceedings of the Hampshire Field Club and Archaeological Society* 26 (1969): fig.29.

75. A. Rogerson and C. Dallas, *Excavations at Thetford 1948–59 and 1973–80,* East Anglian Archaeology 22 (Dereham: Norfolk Archaeological Unit, 1984), 117, 167, 170, 179.

76. Jackson et al, "The Iron Age and Anglo-Saxon site at Upton, Northamptonshire," 210–12.

77. Crummy et al, *Small Finds from the Suburbs and City Defences.*

78. G. Beresford, *Goltho: The Development of an Early Medieval Manor c. 850–1150,* English Heritage Archaeological Report 4 (London: English Heritage, 1987), 3.

79. P. A. Rahtz, *The Saxon and medieval palaces at Cheddar,* British Archaeological Reports British Series 65 (Oxford, 1979), 14–15.

80. Beresford, *Goltho: The Development of an Early Medieval Manor,* 29–30.

81. Beresford, *Goltho: The Development of an Early Medieval Manor,* 55.

82. Beresford, *Goltho: The Development of an Early Medieval Manor,* 55–7, 68; C. A. R. Radford, "The Saxon house: a review and some parallels," *Medieval Archaeology* 1 (1957): 37.

83. R. T. Jones and I. Ruben, "Animal bones, with some notes on the effects of differential sampling" in Beresford, *Goltho: The Development of an Early Medieval Manor,* 197–206.

84. R. Hall, *The Viking Dig* (London: The Bodley Head, 1984), fig.4.

85. R. Hall, "Archaeological introduction" in Walton, *Textiles, Cordage and Raw Fibre,* 294.

86. Walton, *Textiles, Cordage and Raw Fibre,* 432–4; Walton Rogers, *Textile Production at 16–22 Coppergate,* 1793.

87. Walton Rogers, *Textile Production at 16–22 Coppergate,* 1797–1801.

88. Hall in Walton, *Textiles, Cordage and Raw Fibre,* 295.

89. Walton, *Textiles, Cordage and Raw Fibre,* 434–7.

90. Walton Rogers, *Textile Production at 16–22 Coppergate,* 1801.

91. Hall in Walton, *Textiles, Cordage and Raw Fibre,* 295.

92. Walton, *Textiles, Cordage and Raw Fibre,* 438–40.

93. Walton, *Textiles, Cordage and Raw Fibre,* 411.

94. Walton, *Textiles, Cordage and Raw Fibre,* 397–403.

95. One bowl fragment; Rogerson and Dallas, *Excavations at Thetford 1948–59 and 1973–80,* 167.

96. Pritchard, "Small Finds," 168–9.

97. Pritchard, "Late Saxon textiles from the City of London," 57–8.

98. Pritchard, "Late Saxon textiles from the City of London," 58.

99. Ottaway, *Anglo-Scandinavian Ironwork from Coppergate*, 538–51; Walton Rogers, *Textile Production at 16–22 Coppergate*.
100. Walton Rogers, *Textile Production at 16–22 Coppergate*, 1824–5.
101. Walton Rogers, *Textile Production at 16–22 Coppergate*, fig.793, fig.856.

CHAPTER 3

DRESSING THE PART:
DEPICTIONS OF NOBLE COSTUME
IN IRISH HIGH CROSSES

Maggie McEnchroe Williams

You Are What You Wear

In the tale of King Niall of the Nine Hostages, progenitor of the powerful Uí Néill dynasty, leadership is bestowed upon the young Niall by a beautiful, finely dressed lady:

> Like the end of snow in trenches was every bit of her from head to sole. Plump and queenly fore-arms she had: fingers long and lengthy: calves straight and beautifully coloured. Two blunt shoes of white bronze between her little, soft-white feet and the ground. A costly full-purple mantle she wore, with a brooch of bright silver in the clothing of the mantle. Shining pearly teeth she had, an eye large and queenly, and lips red as rowanberries. . . .
> "Who art thou?" says the boy. "I am the Sovranty," she answered.[1]

This personification of leadership is not only presented as a beautiful creature, but also as a wealthy noblewoman who is exquisitely dressed. Her costume includes elaborately woven textiles and a precious brooch that effectively prove her privileged social standing, wealth, and political prominence. Although she initially appears to Niall as a hideous hag, once he lies with her, she is transformed into the stunning image of kingship described above. Her metamorphosis shows that she is a powerful shapeshifter, while her chosen identity as a comely lady in a fine purple mantle confirms that she has earthly powers as well. In fact, her physical attractiveness and expensive garments serve as immediate signals of her social and financial status, and those same visual attributes legitimate her ability to dole out power. This mystical figure of "Sovranty" displays both the physical perfection and the expensive attire that were the hallmarks of early Irish kings, queens, and warriors.[2]

The precise depiction of certain types of attire on the sculptures known as high crosses served to construct and perpetuate a particular notion of authoritative costume in medieval Ireland. The crosses are enormous artifacts—some of which are nearly 20 feet high—and they are characterized by a ring encircling the intersection of their shafts and arms. Their surfaces are adorned with relief sculptures including elaborately carved geometric designs as well as sophisticated figural compositions. I have chosen several relief carvings that appear on three of the crosses—at Clonmacnois, county Offaly; Monasterboice, County Louth; and Kells, County Meath. In each relief, male figures appear wearing a specific style of dress that consists of an amalgam of ancient and medieval fashions. Their clothing serves to define their identities as Irish kings and noblemen, establishing a firm visual code for upper-class male Irish costume. [3]

This notion of selecting a certain style of dress and fixing it in a particular state in order to serve as a definitive sign of a culture's chosen identity is not unique. For example, the Scottish kilt serves an analogous function. The modern kilt is a familiar indicator of a particular heritage, and it is generally thought to have derived from an ancient fashion. However, as Malcolm Chapman has demonstrated, the kilt was only adopted as a "traditional" style of dress in the late eighteenth century. Chapman concludes that the reconstruction of this antiquated costume from a variety of sources had to do with a contemporary eighteenth-century political need for self-definition, and that it became most effective in that capacity within the tourist industry.[4] As Chapman shows, Scottish Highland dress was devised in order to define a cultural and political identity that contrasted native Scotsmen with non-Scottish outsiders; it is a costume that was actively and intentionally created to embody a particular cultural "tradition."[5]

Similarly, the carvings of noblemen and kings on the Irish high crosses portray a specific, codified language of dress that signifies the individuals' cultural, political, and class identities. In fact, as I will demonstrate, the carvings depict a rigidly defined costume that is used to signal a particular class of Irish man.[6] Moreover, the sculptures depict garments that fit descriptions in texts whose dates span thousands of years, suggesting that the attire is not representative of a fleeting trend or a contemporary fashion. The connection between style of dress and an Irish noble identity was permanently and publicly advocated on the crosses.

The Léine and Brat

Both written descriptions and visual depictions of Irish garments in the early Middle Ages present a particular style of clothing more often than any other: the *léine* and *brat*. The *léine* was a sleeveless smock-like garment of variable length that was sometimes hooded. It could be made of un-bleached or white linen—or silk on rare occasions—and sometimes had a decorated hem.[7] This light undergarment was covered by a *brat*, a brightly-colored, four-cornered shawl made of wool that could be wrapped around the body and fastened with a brooch.[8] Although there is no way to know for certain whether or not medieval Irish people actually wore this type of costume, cultural products ranging from literary texts and visual images to legal tracts present it as an indispensable marker of the Irish nobility.

For instance, the epic tale *Táin Bó Cúalnge* [*The Cattle Raid of Cooley*] describes elaborately decorated examples of the *léine* and *brat*. In the *Táin*, the fiercest Irish warriors battle one another for possession of the Brown Bull of Cúalnge, believed to be the finest animal in the country. Cattle were valuable commodities in early medieval Ireland and raids on others' herds were a common occurrence. In this text, the *léine* and *brat* are ascribed to Bronze Age warrior-kings; however, most scholars agree that the story was recited orally long before being transcribed in the eighth or ninth century, and the text only survives in manuscripts of twelfth-century date.[9] Conse-quently, although the content of the narrative is believed to be primarily pre-Christian, certain details such as costume may resonate with an early medieval sensibility as much as with Bronze Age tastes.[10] The text illustrates that the *léine* and *brat* was a fashion that survived for thousands of years, be-coming increasingly venerable with age. By the ninth or tenth centuries, when the crosses were carved, the *léine* and *brat* had become firmly estab-lished as the "traditional" dress of the ancient Irish warrior-kings.

Several passages in *Táin Bó Cúalnge* are devoted to describing the elab-orate attire of the noblest Irish warriors arrayed in their most terrifying and impressive garb.[11] On the night before the final conflict of the tale, twenty noble warriors in elegant costumes lead companies of some three thousand men each onto the battlefield. A scout describes their leader, Conchobor, king of the province of Ulster:

> Finest of the princes of the world was he among his troops, in fearsome-ness and horror, in battle and in contention. Fair yellow hair he had, curled, well-arranged, ringletted, cut short. His countenance was comely and clear crimson. An eager grey eye in his head, fierce and awe-inspiring. A forked beard, yellow and curly, on his chin. A purple mantle fringed, five-folded, about him (*fúan corcra corrtharach cáeicdiabuil imbi*) and a golden brooch in the

mantle over his breast. A pure white, hooded shirt with insertion of red gold he wore next to his white skin. (*Léine glégel chulpatach ba dergintluid do dergór fria gelchness.*) He carried a white shield ornamented with animal designs in red gold. In one hand he had a gold-hilted, ornamented sword, in the other a broad, grey spear. That warrior took up position at the top of the hill and everyone came to him and his company took their places around him.[12]

Significantly, the subordinate groups of soldiers are differentiated from one another by their clothing:

> Another company came . . . second only to the first in numbers and discipline and dress and terrible fierceness. (*Tánaise dá séitche eter lín & chostud & timthaige, Láech cáem cendlethan i n-airinuch na buidnisin.*) A fair young hero headed this company, with a green cloak wrapped around him, fastened at his shoulder with a gold brooch. His hair was curled and yellow. He wore at his left an ivory-hilted sword, the hilt cut from a boar's tusk. A bordered tunic covered him to the knee.[13]

Each squadron leader wears the *léine* and *brat,* and their relative status can only be determined by the quality and color of their costumes. Both written and archaeological evidence confirm that Irish clothing could be dyed many different colors, the most common shades being a dark-brownish yellow, purple or crimson, and green. Mention is also made of black, grey, brown, variegated, and striped garments.[14] In the seventh- or eighth-century Old Irish law-tract, *Senchus Mór,* a statute appears that associates these colors with rank:

> According to the rank of each man, from the humblest man to the king, is the clothing of his son. Blay-colored, and yellow, and black, and white clothes are to be worn by the sons of inferior grades; red, and green, and brown clothes by the sons of chieftains; purple and blue clothes by the sons of kings.[15]

The author of this text points out that this system was not always legally monitored, but was nonetheless customary:

> No book mentions a difference of raiment, or that there should be any difference in their clothes at all. But the custom now is as follows: —Satin and scarlet are for the son of the king of Erin, and silver on his scabbards, and brass rings upon his hurling sticks, and tin upon the scabbards of the sons of chieftains of lower rank, and brass rings upon their hurling-sticks.[16]

The colorful details of such fine costumes may have originally been painted onto the carvings, although no traces of polychromy survive.[17]

The splendor of a nobleman's *léine* and *brat* could be further enhanced through the addition of embroidered details. The exquisite quality of Irish embroidery is well-known in this period, as demonstrated by a reference to its value in the Brehon Laws: "For ornamental work, there is paid to the amount of value of an ounce of silver, for every woman who is an embroideress deserves more profit than even queens."[18]

Jewelry was also a badge of status and wealth. In many of the carvings that adorn the crosses, a man's *brat* is held in place by a large (pen)annular brooch on the right shoulder.[19] (See Figure 3.2 and Figure 3.3.) Many examples of this type of extravagant golden and jeweled Irish brooch have survived—such as the Tara Brooch of ca. 750—and written and archaeological evidence suggest that the size and intricacy of a person's brooch was commensurate with his rank. The *Senchus Mór* specifies that:

> brooches of gold, having crystal inserted in them, with the sons of the king of Erin, and of the king of a province, and brooches of silver with the sons of the king of a territory, or a great territory; or the son of each king is to have a similar brooch, as to material; but that the ornamentation of all these should appear in that brooch.[20]

Such fine brooches, when used to hold a *brat* in place, were sufficient to indicate a nobleman's elevated status, but written sources like *Senchus Mór* and *Táin Bó Cúalnge* diverge on the precise identification of particular ranks. Indeed, it is unlikely that any sumptuary legislation would have been consistently observed from the Bronze Age through the early Middle Ages, and it is reasonable to assume that custom was often victorious over law.

Even though the *léine* and *brat* was not an exclusively Irish fashion—in fact, it closely resembled the Roman *tunica* and *sagum*—it certainly differed from the early tenth-century norm in the rest of Europe.[21] Around the time that many of the Irish crosses were erected, most northern Europeans—including the Scandinavians who had become a permanent fixture in Irish society—were wearing a type of costume that consisted of a short tunic and breeches. Such attire allowed freedom of movement, provided warmth in colder climes, and may have been worn by the lower classes in Ireland; by contrast, the *léine* and *brat* may have been rather impractical as the amount of protection they provided from the cool, blustery Irish weather is debatable. Moreover, such long, heavy clothes must have mitigated any vigorous physical activity.

Indeed, it is possible that medieval Irish people were wearing garments that were more practical than the *léine* and *brat*, but other types of dress appear less frequently in visual images and written sources. One style of clothing that is sometimes mentioned or illustrated is known as the jacket and *truibhas* or *triús*.[22] Undoubtedly suited for active lifestyles, the jacket and *triús* consisted of a short tunic, like a modern shirt, and a pair of breeches or leggings. The written sources do not refer to this costume as frequently as they do to the *léine* and *brat,* presumably because it was worn primarily by the lower classes. The compound term *bern-bróc*—an older synonym for *triús*—is often found in connection with men in subordinate positions, such as charioteers, troops in the king's bodyguard, food-bearers, doorkeepers, and scouts or spies.[23] In addition, a passage in the thirteenth-century *Sagas of the Norse Kings* by Snorre Sturlason describes this type of costume as worn by Harald Gille, a visiting Irishmen: "When Harald came he was dressed thus. He had on a shirt and trousers which were bound with ribands under his foot-soles, a short cloak, an Irish hat on his head, and a spear-shaft in his hand."[24]

Dressing the Part

Carvings depicting the *léine* and *brat* appear on the *Cross of the Scriptures—Cros na Screaptra* in Old Irish—which stands amid the ruins of the early Christian monastery of Clonmacnois, County Offaly.[25] (Figure 3.1.) The cross is a magnificent monolithic structure made from a single block of sandstone. The completed sculpture takes the form of a massive Latin cross with a prominent ring encircling the intersection of its shaft and arms, and it measures almost four meters in height. Every surface is adorned with relief carvings including figural scenes and panels of intricate geometric designs. Depictions of *Christ in the Tomb, Christ's Arrest,* and *The Rending of Christ's Garments* occupy the west face, while the *Traditio Clavium* appears above two previously unidentified scenes on the east. The narrow north and south sides of the cross include a variety of themes, among them *David Playing his Lyre, Saint Michael Slaying the Dragon,* and *A Cleric Plaiting or Cutting Another's Hair.* The cross's head is decorated with a *Crucifixion* on the west side and a *Last Judgement* on the east, and the base is carved with scenes of chariots and animals. An inscription has been used to date the monument to the early tenth century.[26]

   Two unidentified scenes appear on the east face of the *Cross of the Scriptures.* Scholars have argued that they illustrate episodes from the Joseph story, hagiographical anecdotes from the *Life of Saint Ciarán,* founder of Clonmacnois, or depictions of contemporary historical individuals.[27] Regardless of their particular iconographical interpretations, these two images include figures wearing elaborate examples of the *léine* and *brat,* suggesting that they are Irishmen who belong to the noble classes. In the lowermost panel, an individual in a long, richly ornamented tunic and cloak turns to

Figure 3.1 Detail, lowermost panel, east face, *The Cross of the Scriptures* [*Cros na Screaptra*], Clonmacnois, County Offaly. Photograph by Margaret Williams.

face a bearded man with a shorter garment and a large, Danish-type sword.[28] (Figure 3.1.) A slender cylindrical object that appears to be a staff or vine divides the scene. Both men grasp the central object and step toward it in a gesture of collaboration. Although the figure on the left wears a *léine* and *brat* that appear to be decorated with needlework and applied gems, his lack of a weapon and the small pouch draped over his back are the attributes of another sector of society. His appearance also contrasts sharply with that of the figure on the right side of the panel, whose shorter tunic, abundant beard, and prominent sword suggest that he is a warrior. This figure may have gathered up his *léine* and belted it, a common style for noblemen engaged in active pursuits.[29]

Figure 3.2   Detail, central panel, east face, *The Cross of the Scriptures* [*Cros na Screaptra*], Clonmacnois, County Offaly. Photograph by Margaret Williams.

Most scholars have identified these two personages as a cleric and a warrior.[30] The left-hand figure's hooded *léine* seems to represent a *caputiis* (*cochal*), a hooded, cowl-like garment described by Gerald of Wales in 1188.[31] Gerald's two visits to Ireland did not bring him to every region of the country, and they centered on religious houses; consequently, the garments he describes were probably worn mainly by men of the church. In addition, the small bag that the figure carries resembles later book satchels like the fifteenth-century example that is now associated with the *Book of Armagh*.[32] And, finally, the figure's cropped hair and beardlessness may be the attributes of a cleric.

In the central panel, two bearded figures wear long, close-fitting tunics and cloaks fastened with prominent ring brooches. (Figure 3.2.) Large swords hang from their belts and they appear to be passing an oblong object between them, a gesture of cooperation similar to the one in the image below.[33] This gesture has been interpreted as representative of sealing a pact. Some scholars have suggested that these two figures are Saint Ciarán and King Diarmait Mac Cerbaill formulating their agreement to found the monastery of Clonmacnois, while others have proposed that they are King Fland Sinna and his former enemy Cathal Mac Conchobair, King of Connacht, who negotiated a peace treaty around 900.[34] In that

Figure 3.3 Detail, *Muiredach's Cross* at Monasterboice [*Mainistir-Buithe*], County Louth. Photograph by Margaret Williams.

year, the *Annals of the Four Masters* report: "A meeting at Ath-Luain [Athlone, near Clonmacnois] between Fland, son of Maelseachlainn, and Cathal, son of Conchobhar; and Cathal came into the house of Fland under the protection of the clergy of Ciarán, so that he was afterwards obedient to the king."[35]

As in the image below, the precise identification of these two warrior kings is subordinated to their gesture of friendship. Their static, frontal poses suggest that they are not currently fulfilling their duties as warriors, despite their preparedness for battle. They are wise and ethical leaders who seem to be involved in the business of political decision-making, but they keep their weapons close at hand in order to demonstrate their potential for military prowess. They may be formulating a peace pact, such as that reached between Fland Sinna and Cathal Mac Conchobair, and their gesture of exchange may symbolize the accord.

Although this relief cannot be conclusively identified as a portrait of specific patrons, it may be a generalizing depiction of members of the same rank or kin group. Their ornate brooches declare their nobility, and their long moustaches, plaited and forked beards, and prominent weapons are the attributes of secular individuals. Moreover, each composition evokes an instance of collaboration, and the decorated cloaks of the figures, prominent weapons, and conspicuous jewelry identify them as high-ranking individuals. Rather than emphasizing the particular identities of the subjects, the sculptors have defined their relationships to one another, highlighting both the inextricable links between the ecclesiastical and lay arenas and the powerful bonds between secular noblemen. These are wealthy and sophisticated men whose loyalty to one another provided the stable scaffold upon which the ecclesiastical community and its environs were constructed. The depiction of fine clothing on the *Cross of the Scriptures* at Clonmacnois acts as a kind of visual analogue of the interactions between lay society and the ecclesiastical world, and the emphasis in these images on "traditional" Irish dress is a central factor in their ability to proclaim the benefits of such relationships.[36]

Another depiction of the *léine* and *brat* appears on *Muiredach's Cross* at Monasterboice (*Mainistir-Buithe*), County Louth.[37] (Figure 3.3) The cross measures just over five meters in height and is dominated by a ring of slightly more than two meters in diameter. It is constructed from three large sandstone blocks fitted together with mortise-and-tenon joints.[38] The monument acquired its nickname from an inscription in Old Irish, which has also been used to date the cross to the early tenth century.[39]

Every surface of *Muiredach's Cross* is adorned with relief sculptures. Most of the carvings are figural scenes, representing biblical events, but some are panels of purely geometric decoration or animal interlace. On the east face of the cross, the four scenes on the shaft have been identified as *Adam and Eve and Cain and Abel, David and Goliath, Moses Smites the Rock,* and *The Adoration of the Magi.* On the west face, the iconography is more difficult to recognize, but most scholars have described the scenes as *The Arrest of Christ, The Raised Christ,* and the *Traditio Clavium.* The narrow north and south sides are primarily adorned with interlace decoration, and the base includes several panels of geometric designs interspersed with fantastic beasts and signs of the zodiac. The head is carved with a *Crucifixion* on the west face and a *Last Judgement* on the east.[40]

The lowermost panel on the west face includes three male figures in clearly differentiated costumes: the central figure wears a long robe and cloak, while the two men that flank him wear short pants and shirts, possibly the jacket and *triús.* (Figure 3.3.) The central figure is male, beardless, and has short wavy hair. He holds a rod or staff in his right hand, and his left arm is held aloft by the figure to his left. His head and feet are bare, and he is clothed in a long, close-fitting tunic with a narrow skirt, evidently a *léine.* The fabric of his *léine* is covered with incised lines that seem to articulate either a woven or embroidered pattern in the textile. He also wears a cloak over his *léine,* probably a *brat.* The spiral designs at the base of the central figure's *brat* may indicate a long garment, cinched at the hem; at the very least, the incised designs demonstrate that his *brat* is adorned. His *brat* is held in place by prominent ring brooch.

Despite his apparent wealth and nobility, the central figure in the Monasterboice panel is clearly not the victor in this violent encounter. Nevertheless, his clothing and demeanor announce his dignity and strength in the midst of the attack. Not only does he lack a visible weapon, indicating that he is not a warrior by trade, but his docile submission, short hair, and beardlessness also differentiate him from his foes. Most scholars have argued that this central figure is either Christ, an ecclesiastic, or a saint; however, he is clothed in the typical costume of the secular nobility. Although the *léine* and *brat* were not dissimilar from ecclesiastical garb of the period, this figure wears a large, visible (pen)annular brooch. The figure's identity is consequently slightly ambiguous: his willing surrender and lack of a weapon suggest that he is an ecclesiastic, but his fine costume is a resounding declaration of his status, wealth, and association with the secular nobility. In fact, the depiction of this individual's attire might not actually reflect whatever garb he wore on a daily basis or even on special occasions; rather it could have signified his elevated rank and Monasterboice's influential political connections.

By contrast, the two side figures in the Monasterboice panel certainly seem to be equipped for battle. Both men hold large swords, and they approach the central figure menacingly, grasping his wrist to hold him in position while pressing a weapon against his vulnerable belly. Despite their threatening demeanor and apparent difference in status, the soldiers in this scene are not without personal ornamentation. The figure on the right wears a simple kite-shaped brooch, examples of which have survived from the eighth and ninth centuries. The kite-shaped brooch was a cheaper and more common alternative to the large, decorated (pen)annular brooches such as that worn by the central figure in this scene, and they were consequently an obvious visual indicator of lower social rank.[41]

Most scholars have identified the scene on *Muiredach's Cross* as either the *Arrest of Christ* or the *Ecce Rex Iudaeorum* or *Second Mocking of Christ*, in which Jesus is dressed in mock-royal robes. Some have suggested that it may represent an ecclesiastic attacked by two armed men, Saint Columcille being arrested and exiled, or Norsemen attacking the abbot of a local community.[42] It is tempting to believe that the images depict historical figures, for that would easily explain the inclusion of familiar Irish costume. If that is the case, then the image represents a high-ranking churchman—whose elaborate garb connects him with the secular nobility—being attacked by lower-class lay warriors.

Even if we agree with those scholars who argue that these scenes represent episodes from the *Passion of Christ*, the inclusion of figures wearing the *léine* and *brat* and jacket and *triús* still warrants attention. If the bottom scene is either the *Arrest of Christ* or the *Ecce Rex Iudaeorum,* then the savior is depicted in the attire of an Irish nobleman, complete with a prominent (pen)annular brooch holding his *brat* in place. His captors wear typical northern European costume, and are branded as inferior by their style of jewelry. This crucial episode in New Testament history is cast as a local social drama, and the significance of all three figures and their actions becomes clear through the details of their dress: two coarse working men, possibly foreigners, brutally attack the regal personage of a religious leader, probably Jesus Christ himself.

My final example derives from the so-called "Market Cross" at Kells, county Meath, which has been dated to the mid-ninth or early tenth century on the basis of stylistic and iconographic similarities with other monuments like the *Cross of the Scriptures* at Clonmacnois and *Muiredach's Cross* at Monasterboice. The cross is made of sandstone, it measures over three meters in height, and it is covered with figural sculpture in relief. The base is decorated with images of horsemen, archers, and animals. On the west face of the shaft,[43] a seventeenth-century inscription occupies the lowermost panel, a space that Helen Roe suggests was originally filled with a depiction of the *Baptism of Christ.*[44] Above that, there are images of the *Adoration of the Magi,* the *Miracle at Cana,* and the *Miracle of the Loaves and Fishes.* On the west face of the cross's head is a carving of the *Crucifixion.*

On the east face, above a panel of spiral designs, is an image of *Christ in the Tomb,* an unidentified scene, and a divided panel with *Adam and Eve* on the left and *Cain and Abel* on the right. *Daniel in the Lions' Den* appears on the head with a depiction of the *Sacrifice of Isaac* on the left arm and an unidentified image on the right arm.

The carving on the right arm shows a standing, frontal figure wearing a long garment and a prominent ring brooch. On either side of this gentleman, there are strange, demonic creatures that clutch at his garments. The creature on the left wears a hood, while that on the right has a horned goat's head. Several scholars, including Arthur Kingsley Porter and Peter Harbison, have identified this scene as the *Temptation of Saint Anthony by Devils.* However, Helen Roe made the intriguing suggestion that this figural group might be an image of the *Deadly Sin of Avarice,* which she compares with Romanesque examples of the figure of a miser carried off by demons.[45]

Whether the central figure can be identified as a personification of Avarice or an image of Saint Anthony, he is clearly dressed in the familiar costume of the Irish nobility, complete with an unmistakable ring brooch.[46] Aggressive demons whisper into his ears, suggesting that temptation lurks nearby, and that it may be targeting wealthy Irishmen in par-

ticular. Moreover, the image seems to resonate with the world of commercial exchange, possibly providing additional evidence for the existence
of some sort of market at Kells.

Conclusion

The images of male figures wearing the *léine* and *brat* at Clonmacnois,
Monasterboice, and Kells perpetuate a specific visual code for noble Irish
dress. Using these particular garments, the designers and carvers of the
crosses were able to promote specific costumes as markers of wealth, status, and power. Moreover, the interactions between the figures that wear
such attire—as well as the encounters with men wearing the costume of
another rank—help to construct a public image of the nobility's role in
Irish society. Noblemen are depicted as people who make pacts with one
another and help to found monasteries, as seen on the *Cross of the Scriptures*
at Clonmacnois. They are shown defending monasteries, and imitating the
virtues of Christ himself, as on *Muiredach's Cross* at Monasterboice. And,
they are encouraged to be mindful of conducting fair and just economic
transactions, lest they be accosted by demons as on the Market Cross at
Kells. Ultimately, these images participated in an ongoing discourse about
the outward appearance of the Irish nobility. They operated in conjunction
with texts and presumably also with contemporary rituals to promote a
particular notion of appropriate noble costume and behavior in early medieval Ireland.

Notes

1. The historical king Niall reigned from ca. 379–*ca.* 405, but this text probably dates to the eleventh century. The translation is Whitley Stokes's, from
   the original, "Ba samalta fri dered snechta i claidib cach n-alt o ind co
   bond di. Rigthi remra rignaidhe lé. Méra seta sithlebra. Colpta dirgi
   dathailli le. Da maelasa findruine iter a troigthib mine maethgela & lar. Brat
   logmarda lancorcra impi. Bretnass gelairgit i timthach in bruit. Fiacla niamda nemannda le, & rosc rignaide romor, & beoil partardeirg . . . 'Cia
   tusu?' or in mac. 'Misi in Flaithius,' or si. . . ." See Whitley Stokes, "The
   Death of Crimthan, Son of Fidach, and the Adventures of the Sons of
   Eochaid Muigmedon," *Revue Celtique* XXIV (1903): 172–207, at 198–201.
   See also Myles Dillon, *The Cycles of the Kings* (Dublin: Four Courts Press,
   1994), 38–40.
2. As several early law tracts demonstrate, Irish kings were expected to be free
   of any visible deformities. See Fergus Kelly, *A Guide to Early Irish Law,* vol.
   3, *Early Irish Law Series* (Dublin: Institute for Advanced Studies, 1988), 19.
   Moreover, in addition to the citations provided below, descriptions of Irish

kings and noblemen in elaborate attire can also be found in Myles Dillon, *The Cycles of the Kings*, 1994; and C. Plummer, *Vitae Sanctorum Hiberniae*, 2 vols. (Oxford: Oxford University Press, 1910).

3. As many scholars have shown, the garments that people wear constitute a nonverbal system of communication, serving as an immediate, visible, and performative language through which one publicly proclaims membership in a particular group, sometimes adding unique touches that represent individual personality traits or choices. For instance, see Malcolm Barnard, *Fashion as Communication* (New York: Routledge, 1996); Roland Barthes, "The Garment System," in *Elements of Semiology*, trans. Annette Lavers and Colin Smith (New York: Hill and Wang, 1968), 25–28; Roland Barthes, *The Fashion System*, trans. Matthew Ward and Richard Howard (New York: Hill and Wang, 1983). On the anthropology of costume, see Georg Simmel, "Philosophie der Mode (1905)," in Georg Simmel, *Gesamtausgabe*, Herausgegeben von Otthein Rammstedt (Frankfurt am Main: Suhrkamp Verlag, 1995), 9–37. In an interesting collection of essays on the topic of ethnic dress, Joanne Eicher and others describe the function of clothing in defining ethnic identity. See Joanne B. Eicher, ed., *Dress and Ethnicity* (Oxford: Berg Publishers Limited, 1995).

4. Malcolm Chapman, "'Freezing the Frame': Dress and Ethnicity in Brittany and Gaelic Scotland," in J. Eicher, ed., *Dress and Ethnicity*, 1995, 7–28, at 15. See also M. Chapman, *The Celts: the Construction of a Myth* (London: Macmillan, 1992).

5. On the issue of creating traditions as a means of defining a culture, see Eric Hobsbawm and Terence Ranger, eds., *The Invention of Tradition* (London: Cambridge University Press, 1983).

6. This notion of costume as a formal, structured system of dress rather than a more personal means of expression has been examined extensively by Roland Barthes. Barthes differentiated between costume (language) and clothing (speech), saying that " . . . clothing always draws on costume . . . but costume . . . *precedes* [Barthes's emphasis] clothing, since it comes from the ready-made industry, that is, from a minority group. . . ." See R. Barthes, "The Garment System," in *Elements of Semiology*, 1968, 27.

7. The term "*léine*" appears quite frequently in the literary sources and the adjective *gel* meaning bright is often used to describe it. See C. O'Rahilly, ed. and trans., *Táin Bó Cúalnge from the Book of Leinster*, vol. 49, *Irish Texts Society* (Dublin: Institute for Advanced Studies, 1967). *Léine* appears to be a native word and may derive from a root meaning linen. See M. Dunlevy, *Dress in Ireland* (New York: Holmes & Meier, 1989), 15–26; F. Shaw, S. J., "Irish Dress in Pre-Norman Times," in *Old Irish and Highland Dress*, 2nd edition, ed. H. F. McClintock (Dundalk: Dundalgan Press, 1950), 12–13.

8. *Brat* also seems to be a native word, although it is of uncertain derivation. Several synonyms for this type of garment appear in the literature including *fuan, lend,* and *lumman*. A *corrthar*—a border or fringe that was woven

separately—was sometimes attached to the *brat*. See F. Shaw, S. J., "Irish Dress in Pre-Norman Times," 1950, 12–13.

9. Thomas Kinsella, ed. and trans., *The Táin: From the Irish Epic Táin Bó Cuailnge* (Oxford: Oxford University Press, 1969), ix–xvi.

10. In particular, the frequent references to silk *léinte* (pl.) suggest an early medieval inflection because silk was probably not readily available in Ireland until the ninth or tenth century. Excavations in Dublin revealed imported compound silks, silk tabbies, and gold braids dating to the tenth century. See Patrick Wallace, "The Archaeology of Viking Dublin," in *The Comparative History of Urban Origins in Non-Roman Europe: Ireland, Wales, Denmark, Germany, Poland and Russia from the ninth to the thirteenth century,* ed. H. B. Clarke and Anngret Simms, British Archaeological Reports International Series (Oxford: B.A.R., 1985), 103–45 at 135.

11. Once a warrior entered into the fray, he was likely to adjust—or even remove—his finery for the duration of the battle. There are references in the *Táin* to soldiers rushing naked into battle, and also to warriors' "battle-harnesses," which apparently consisted of leather protective gear. See Thomas Kinsella, ed. and trans., *The Táin,* 1969, 243. Classical authors also described the Celts fighting in the nude. In the first century B.C.E., Diodorus Siculus wrote, "Some of them have iron cuirasses, chain-wrought, but others are satisfied with the armour which Nature has given them and go into battle naked." See C. H. Oldfather, trans., *The Library of History of Diodorus of Sicily,* ed. E. H. Warmington, vol. 3 of 12, *Loeb Classical Library* (Cambridge, Mass.: Harvard University Press, 1970 [1939]), 177.

12. C. O'Rahilly, ed. and trans., *Táin Bó Cúalnge from the Book of Leinster,* 1967, 119, 254.

13. I am using Thomas Kinsella's translation here because he includes an opening phrase in which dress is specifically listed as a criterion for distinguishing between the ranks. Thomas Kinsella, ed. and trans., *The Táin,* 1969, 226. In C. O'Rahilly's version, the distinction between costumes is certainly implied, but it is not stated explicitly. O'Rahilly translates the same passage as follows: "A handsome man in the forefront of that same band. Fair yellow hair he had. A bright and very curly beard on his chin. A green mantle wrapt around him. A pure silver brooch in the mantle over his breast. A dark-red, soldierly tunic with insertion of red gold next to his fair skin and reaching to his knees." C. O'Rahilly, ed. and trans., *Táin Bó Cúalnge from the Book of Leinster,* 1967, 119, 254–5.

14. Yellow seems to have been a very popular color, which the Irish may have achieved by using a local plant called *buidh mor* (great yellow) rather than imported saffron. See J. C. Walker, *An Historical Essay on the Dress of the Ancient and Modern Irish,* (Dublin: J. Christie, 1818), 262. Purple dye also seems to have been made in Ireland. See F. Henry, "A Wooden Hut on Inishkea North, Co. Mayo," *Journal of the Royal Society of Antiquaries of Ireland,* vol. 52/2 (1952): 163–178. Additional evidence for brightly-colored costumes comes from the illuminated manuscripts of the period, particularly the

Book of Kells, Dublin (TCD MS 1), and also from the archaeological textiles found at Lagore crannog, which might have been dyed red. See L. Start, "Textiles," in "Lagore crannog: an Irish royal residence of the seventh to the ninth centuries" H. O'Neill Hencken, *Proceedings of the Royal Irish Academy,* vol. 53C (1950–1), 1–247 at 214.

15. W. N. Hancock and T. O'Mahony, eds., *Ancient Laws of Ireland,* 6 vols. (Dublin: HMSO, 1865–1901), vol. 2, 147–9.

16. W. N. Hancock and T. O'Mahony, eds., *Ancient Laws of Ireland,* vol. 2, 147–9.

17. Traces of polychromy survive on continental sculptures, and it is reasonable to assume that Irish sculpture could also have been painted in this period, particularly considering the importance of color in determining a person's rank and identity. Ireland's damp climate is probably responsible for the erosion of any traces of paint.

18. Hancock and O'Mahony, vol. 5, 383.

19. The condition of the carvings makes it difficult to determine whether these are annular or penannular brooches. Both types are circular, but the penannular variety has a gap in the ring. See Susan Youngs, ed., *The Work of Angels: Masterpieces of Celtic Metalwork, 6th–9th Centuries AD* (Austin: University of Texas Press, 1989), 214–215.

20. Hancock and O'Mahony, vol. 2, 147–9.

21. It is possible that the *léine* and *brat* may have gained popularity as an imitation of Roman dress. When writing in Latin, Irish authors often used the word *tunica* to describe a *léine*-like garment. See F. Shaw, S. J., "Irish Dress in Pre-Norman Times," 1950, 11–18.

22. The word *tríubhas* or *triús* appears to be derived from the Old French *trebus,* which is also the origin of the English word *trousers.* See F. Shaw, S. J., "Irish Dress in Pre-Norman Times," 1950, 16–17. J. C. Walker used the term *cota* to describe a shirt that fell to the loins, probably the so-called jacket. See J. C. Walker, *An Historical Essay on the Dress of the Ancient and Modern Irish,* 1818, 9. Figures wearing this type of costume appear in the *Book of Kells* (TDC ms1) on folios 200R and 130R and on the twelfth-century Aghadoe Crozier.

23. Before 1200, *triús* were probably called *bróc* or *bern-bróc,* which Kuno Meyer defines as "breeches, long hose, or trousers." See Kuno Meyer, *Contributions to Irish Lexicography* (London: D. Nutt, 1906). On occurrences of the term *bern-bróc,* see H. Zimmer, *Zeitschrift für vergleichende Sprachforschung,* vol. XXX (Gütersloh: C. Bertelsmann, 1888). *Triús* were also known among the Romans by their Gaulish name, *bracae* or *braccae.* See F. Shaw, S. J., "Irish Dress in Pre-Norman Times," 1950, 16–17.

24. *The Heimskringla or The Sagas of the Norse Kings from the Icelandic of Snorre Sturlason,* trans. S. Laing, 2 ed. (London: John C. Nimmo, 1898), 168–9.

25. The *Cross of the Scriptures* takes its name from an eleventh-century entry in the annals. See D. Murphy, ed., *The Annals of Clonmacnois, being the Annals of Ireland from the earliest period to A.D. 1408* (Dublin: Royal Society of Antiquar-

ies of Ireland, 1896), 1060 c.e. 178; J. O' Donovan, ed., *Annala Rioghachta Eireann, Annals of the Kingdom of Ireland by the Four Masters*, 6 vols. (Dublin: Hodges, Smith and Co., 1856) [Hereafter AFM], 1060 c.e., vol. 2, 879.

26. Beginning on the west face and continuing on the east, the inscription has been reconstructed as: ORDORIGFL.IND MMA/N/ROIT-DORIGHERENNOR [Pray for Fland, son of Maelsechnaill, prayer for the king of Ireland; prayer. . . .] DOCOLMANDORRO/AN-CROSSAAR/RIGFL.ND [. . . for Colman, who made this cross for King Fland]. King Flann or Fland Sinna reigned from 877 to 914 c.e. While the inscription clearly designates Fland as the person to whom the cross is dedicated, Colman's role is less certain: he may have been a co-patron, designer, or perhaps even sculptor. For more on Colman's position, see Douglas Mac Lean, "The Status of the Sculptor in Old-Irish Law and the Evidence of the Crosses," *Peritia* 9 (1995): 125–55. Colman has been identified as Abbot Colman Conaillech (d. *ca.*921), a contemporary of King Fland's, a discovery that provides an early tenth-century date for the cross. See Peter Harbison, *The High Crosses of Ireland: An Iconographical and Photographic Survey*, 3 vols. (Bonn: R. Habelt, 1992), 48–53.

27. The lowermost scene has been alternately identified as Joseph Interpreting the Dream of Pharaoh's Butler, Moses and the Brazen Serpent, Adam and Eve, King Fland Sinna and Abbot Colman Conaillech building a new stone church in the early tenth century, or Diarmait Mac Cerbaill and Saint Ciarán founding the monastery in the sixth century. The two standing figures in the center panel have been identified as the Chief Butler giving the Cup into Pharaoh's Hand, Dermot and Mael-Mor, Saint Ciarán and Diarmait Mac Cerbaill founding the monastery, and King Fland and Cathal Mac Conchobair or unknown noblemen or chieftains forging an alliance. For a complete description of these interpretations, see P. Harbison, *The High Crosses of Ireland*, 1992, 49. I have argued elsewhere that these two carvings, like many medieval images, resonate on a number of levels, incorporating references to the Bible, the lives of the saints, and contemporary politics. In fact, the figures's costumes trigger a range of associations that enrich the cross's significance by linking the monastic brethren with the Irish nobility. See Williams, Margaret M. "Warrior Kings and Savvy Abbots: The Sacred, The Secular, and the Depiction of Contemporary Costume on the Cross of the Scriptures, Clonmacnois." *Avista Forum Journal* 12, 1 (1999): 4–11.

28. Viking swords were larger and stronger than Irish weapons, and the Irish adopted Viking arms from an early date in order to combat the invaders more effectively. See J. Graham-Campbell and D. Kidd, *The Vikings* (London: British Museum, 1980), 113–4; L. and M. De Paor, *Early Christian Ireland* (London: Thames & Hudson, 1958), 105.

29. According to the late twelfth-century parody of the popular vision-tale genre *Aislinge Meic Conglinne* [*The Vision of Mac Conglinne*], Aniér Mac

Conglinne, a famous scholar, pulled his long *léine* up over his belt in order to prepare for his walk from Roscommon to Cork. See Kuno Meyer, ed. and trans., *Aislinge Meic Conglinne: The Vision of Mac Conglinne* (London: David Nutt, 1892). The carving might also depict a shorter tunic called a *léinte* or *leinidh,* but the extreme variation in Old Irish orthography suggests that all three words are simply alternate spellings for the same garment. See H. F. Mc Clintock, *Old Irish and Highland Dress,* 1943, 121.

30. For a discussion of the many iconographic interpretations, see P. Harbison, *The High Crosses of Ireland,* 1992, 49.

31. John O'Meara, trans., *Gerald of Wales: The History and Topography of Ireland* (London: Penguin Books, 1982 (1951)), 101. On the *cochal,* see J. C. Walker, *An historical essay on the dress of the ancient and modern Irish,* 1818, 12–13. This figure might also be wearing a type of footwear called "sole-less stockings." See J. W. Barber, "Some Observations on Early Christian Footwear," *Journal of the County Kildare Archaeological Society* 86, 243 (1981): 103–106; A. T. Lucas, "Footwear in Ireland," *The Journal of the County Louth Archaeological Society,* 13, 4 (1956): 309–94.

32. Although the satchel has been linked with the *Book of Armagh,* it probably originally held a larger object. See M. Ryan, ed., *Treasures of Ireland: Irish Art 3000 B.C.–1500 A.D.* (Dublin: Royal Irish Academy, 1983), 178–9.

33. Carol Neuman de Vegvar has suggested that the object could be a drinking horn, "Drinking Horns and Social Discourse in Early Medieval Britain and Ireland," (New York: Columbia University's Medieval Seminar, 1997).

34. P. Harbison, *The High Crosses of Ireland,* 1992, 49.

35. AFM, vol. 1, 900 C.E., 554.

36. For more on the links between the sacred and secular realms in early medieval Ireland, see my doctoral dissertation, Margaret McEnchroe Williams, "The Sign of the Cross: Irish High Crosses as Cultural Emblems" (Ph.D., Columbia University, 2000). See also Lisa M. Bitel, *Isle of the Saints: Monastic Settlement and Christian Community in Early Ireland* (Ithaca: Cornell University Press, 1990); Catherine Herbert, "Psalms in Stone: Royalty and Spirituality on Irish High Crosses" (Ph.D. Dissertation, University of Delaware, 1997), 273, 287.

37. For more on the depictions of costume on Muiredach's cross, see Maggie McEnchroe Williams, "And They Clothed Him in Purple: Dressing the Church in Royal Robes at Monasterboice, County Louth" (forthcoming).

38. Roger Stalley, *Irish High Crosses* (Dublin: Eason & Son Ltd., 1991), 2.

39. The inscription reads: OR DO MUIREDACH LAS NDERN(A)D (I) CROS(SSA) [Prayer for Muiredach who had the cross erected.] The annals refer to two individuals named Muiredach, both of whom were abbots at Monasterboice: the first, Muireadhach mac Flaind, held his post from 837 to 846, and the second, Muireadhach mac Domhnaill, from ca. 887 to 922. The latter was simultaneously abbot-elect of the powerful monastery at Armagh, as well as High Steward of the Uí Néill family. As a result of his prominent political affiliations, this second Muiredach has gen-

erally been associated with the cross's inscription, providing a date for the monument of ca. 922–3. See P. Harbison, *The High Crosses of Ireland*, 1992, 364; and Helen M. Roe, *Monasterboice and its Monuments*, 1981, 9.

40. For more on the cross's iconography, see P. Harbison, *The High Crosses of Ireland*, 1992, 140–6.

41. O. Somerville, "Kite-Shaped Brooches," *Journal of the Royal Society of Antiquaries of Ireland* vol. 123 (1993): 59–101; Niamh Whitfield, "The Kite Brooch as Indicator of Social Change in Ireland from the Ninth to the Twelfth Centuries," (Kalamazoo, MI: 33rd International Congress on Medieval Studies, 1998).

42. See P. Harbison, *The High Crosses of Ireland*, 1992, 143; A. K. Porter, *The Crosses and Culture of Ireland* (New Haven: Yale University Press, 1931), 42; E. H. L. Sexton, *A Descriptive & Bibliographical List of Irish Figure Sculptures of the Early Christian Period* (Portland: The Southworth-Anthoensen Press, 1946), 232.

43. The fact that the Kells cross has been moved at least once makes it impossible to tell which way it was originally facing. However, it is likely that the face upon which the *Crucifixion* is carved was the original west face, as is the case with so many other crosses that are still in situ.

44. The English inscription explains that the Kells cross was re-erected in the seventeenth century. It reads: THIS CROSS/WAS ERECTED/(AT) THE CHAR/GE OF ROBERT/(BA)LFE OF GALL/IRSTOWNE ES (Q)/(BEI)NG SOVERAI/(GN)E OF THE CORP/ORATION OF KEL(L)/IS. ANNO DOMI/1688. See Helen Roe, *The High Crosses of Kells* (Kells: Meath Archaeological and Historical Society, 1959), 27–30, 35. For a discussion of the entire cross's iconography, see P. Harbison, *The High Crosses of Ireland*, 1992, 103–8.

45. H. Roe, *The High Crosses of Kells*, 1959, 30–1.

46. Often, continental Romanesque images of Avarice include a figure carrying a large sack of money, which is usually hanging around the individual's neck. Perhaps the ring brooch worn by the figure in the Kells carving serves a similar iconographic function, providing immediate visual evidence of the man's exorbitant wealth.

PART TWO

THE CENTRAL MIDDLE AGES

CHAPTER 4

MARIE DE FRANCE'S *BISCLAVRET:*
WHAT THE WEREWOLF WILL
AND WILL NOT WEAR

Gloria Thomas Gilmore

T his chapter will examine the role that textiles, here specifically clothing, play in the spinning out, and tangling up, of the *Lais* of Marie de
France.[1] Within the texts of these love stories, the *Nightingale's* shroud, as
well as *Fresne's* baby quilt clearly function as subtexts, as "écriture fémi-
nine," or women's writing, because textile work has been the primary responsibility of women for millennia.[2] In these stories, textiles are
deciphered by characters and readers alike. An examination of all references
to textiles reveals that many of them comment on the theme of subject
formation, as do Hanning and Ferrante in the introduction to their modern English translation to the *Lais.* They identify as

> one of the themes explored in 12th century courtly narrative . . . the indi-
> vidual's recognition of a need for self-fulfillment and his or her struggle for
> the freedom to satisfy that need. The tension between the personal quest . . .
> and one's social obligations was a recurring theme of courtly literature. . . .

This chapter will attempt to unravel that tangled tension in the story of
Bisclavret, where there are two opposing functions of clothing: to confine
in a social role or identity imposed from without, or to express a self-
definition, chosen or generated from within. The textiles highlight the necessity of free choice and balance.

Bisclavret, a term for werewolf from either *bleiz lavaret* (the speaking
wolf) or *bisc lavret* (the wolf who wears pants),[3] is the only name given for
the hero, a nobleman who goes deep into the forest to undress and roam

as a werewolf for three days each week. His wife worries about his absences and coaxes him to divulge where he goes, whether he is dressed or naked, and finally where he hides his clothing. He admits that he could never become a man again if his clothing were lost. She wants nothing more to do with him and offers herself to a knight who had been suing for her love (*Bisclavret* 103–5). She then sends her lover to get the clothing, and Bisclavret is trapped in the body of a werewolf. Because he has disappeared so frequently in the past, everyone assumes he has simply gone away forever, instead of for his usual three days. His wife is then free to wed her lover. A year passes. Then the King and his hunting party come across Bisclavret in the forest. The werewolf runs to "beg mercy" *(Bisclavret* 146) from the King by kissing his foot, the King spares his life, and takes him back to the castle with him. Bisclavret loyally accompanies the King wherever he goes and is fondly cared for by all the knights.

When the King summons his barons to court, Bisclavret recognizes his wife's new husband, and attacks him. All assume he has good reason to do so, however. The King then goes hunting in the forest again, and the wife seeks him out to bring him gifts. Bisclavret attacks her and bites off her nose.[4] Recognizing her as the wife of the vanished baron whom the King had long esteemed, a wise man urges him to torture her to find out why Bisclavret would attack only her and her new husband. She tells all, sends for the clothing on command, and is banished with her "husband." Her girl descendants are born noseless, as the self-perpetuating mark of the adulterous relationship that bore them, as well as the reason for their exiled wanderings.[5] Bisclavret shyly ignores the clothing until it is placed in the King's quarters and he is allowed to dress in private. The King finds him sleeping, in human form, on his bed, and rushes to hug and kiss him more than a hundred times, then "Plus li duna ke jeo ne di [Gave him more than I can tell.]" *(Bisclavret* 304).

An examination of three word pairs, and attempting to determine how they affect or reflect the hero's subjectivity will illustrate the textile's confinement of the hero in a social role or identity imposed from without, or its expression of a self-definition, chosen or generated from within.[6] The upper-case spellings of Bisclavret, not random at all, are used consistently to portray the human hero in some social role or context, most often relating to his wife. That the parameters of Bisclavret's social existence should be thus defined in terms of his wife is key, as well as consistent with the unfolding of the plot: it is she who comes to control the pivotal point of Bisclavret's exit and re-entry into human form by taking possession of his clothing. A listing of all the occurrences of the term "Bisclavret" will demonstrate how they illustrate social aspects of the hero's life. First of all, the upper-case spelling simply names the hero known in legend that Marie does not want to forget and whom she wishes to talk about:

Ne voil ublier Bisclavret; . . .
Del Bisclavret vus voil cunter.

[I don't want to forget Bisclavret; . . . I want to tell you about the Bisclavret.]
(*Bisclavret* 2, 14).[7]

The social context here is extra-textual; it is Marie's readership and those
who will hear her *Lais* recited aloud whom she addresses. It is with this society at large that she wishes to establish a lasting relationship with the person of her character, Bisclavret.

It is the hero's social contract with his wife that is violated when his
clothes are stolen:

Issi fu Bisclavrert trahiz
E par sa femme maubailiz.

[So Bisclavret was betrayed, ruined by his own wife.] (*Bisclavret* 125–6).

Not only is the public troth betrayed, but her private protestation of love
and fidelity is equally invalidated. The next lines define Bisclavret's wife socially by relating her to him: it is his name that is the possessive noun used
to identify her.

Ki la femme Bisclavret ot . . .
La femme Bisclavret le sot

[The one who had married Bisclavret's wife . . . the wife of Bisclavret heard
about it.] (*Bisclavret* 191, 227).

Although the woman is here designated socially as Bisclavret's possession,
we soon find that it is she who will determine Bisclavret's being, through
possession of his clothing.

The proper name then associates Bisclavret's wife with the King, the
two people with whom Bisclavret has most specific public and, at the same
time, intimate, social contracts, which is to say, marriage vows and feudal
oaths of homage:[8]

El demain vait al rei parler,
Riche present li fait porter.
Quant Bisclavret la veit venir, . . .

[The next day [she] went to speak with the King, bringing rich presents for him.
When Bisclavret saw her coming, . . .] *(Bisclavret* 229–31).

Clearly the wife believes that her husband is indeed the werewolf in the King's company:

> Tres bien quidot e bien creeit
> Que la beste Bisclavret seit

> [She was quite certain that this beast was Bisclavret.] (*Bisclavret* 273–4).

Although in the end the wife refers to her husband as the "beste," it is her telling of his moving from one form to the other, and of how she trapped him in the one, that brings the proper noun, the concept of the human hero with his social relationships, into public discussion of the activities of the beast. Finally Marie tells us that he, the person, should be remembered forever by all in the society of readers or listeners who hear her lai about him:

> De Bisclavret fu fez li lais
> Pur remembrance a tuz dis mais

> [The lai of Bisclavret was made so it [*he*, author's translation] would be remembered forever.] (*Bisclavret* 317–8).

Her lai is not a gory story about a ghoulish fiend. It is about a social, human character struggling with personal desires, and Marie tells us we should all remember how he nobly resolves his struggles. All of the social settings or relationships described above differ from the situations in which the *bisclavret* appears spelled in lower case.

All of the lower-case spellings, identified below, now refer more often to the private life the hero leads as the wild beast in the forest, than to the socialized man identified by a capitalized Christian name. He is a *bisclavret* in his personal confession of his animal state: "'Dame, jeo deveinc bisclavret' [My dear, I become a werewolf]" (*Bisclavret* 63). And we may note that, except for the preceding exception, an article earmarks him as a savage animal in the forest. First of all, the King goes to the forest, to where the werewolf roams:

> . . . li reis ala chacier.
> A la forest ala tut dreit,
> La u li bisclarvet esteit"

> [. . . the King went hunting, right to the forest where the werewolf was] (author's translation) (*Bisclavret* 136–8).[9]

There the King's dogs find and chase him as a wild beast:

Quant li chien furent descuplé
Le bisclavret unt encuntré

[When the hounds were unleashed, they ran across the werewolf) (author's
translation)
(*Bisclavret* 139–40).[10]

Then at the court gathering he appears and behaves as the wild animal attacking the new husband:

Si tost cum il vint al paleis
E li bisclavret l'aperceut,
De plain esleis vers lui curut:
As danz le prist, vers lui le trait.

[As soon as he came to the palace the werewolf saw him, ran toward him at
full speed, sank his teeth into him, and started to drag him down.] (author's
translation) (*Bisclavret* 196–9).

Of course only a wild beast would try to devour a human. The werewolf, whose savage passion is described as what drives the "garwalf" to
devour men, understandably hates the knight who stole not only his
wife, but also the clothing necessary for him to transform back into
human form:

Alez s'en est li chevaliers
Mien esciént tut as premiers,
Que li bisclavret asailli.
N'est merveille s'il le haï!

[The very first to leave, to the best of my knowledge, was the knight whom
the werewolf had attacked (author's translation). It's no wonder the creature
hated him!] (*Bisclavret* 215–18).[11]

Then the King returns to the forest where he first found the wild animal:

. . . a la forest ala li reis,
U li bisclavret fu trovez.

[. . . the king went to the forest where the werewolf was found.] (author's
translation) (*Bisclavret* 221, 223).[12]

Finally the clothes are given to the suddenly shy beast: "Al bisclavret la fist
doner. [He had them brought to the werewolf.]" (author's translation)

(*Bisclavret* 278).[13] In each situation the "bisclavret," accompanied by an article, invariably speaks of the hidden life of the werewolf, not of the man in any human context.[14]

The split in Bisclavret's state of being has been easily identified as the universal duality of man's nature: the savage, self-indulgent, even sinful core hidden in each of us, which is often only masked by the civilized, socialized, even saved or redeemed veneer presented to the world. However such a dichotomy is seriously put in question by the events and characters of this story. First of all, why would one consider the simple beast of nature "self-indulgent?" It is known that the clothing is necessary for the hero to return to human form. Reasoning backwards, we may assume that it is the clothing that keeps him from transforming into a werewolf. Hence, it must be a conscious decision on the human hero's part to remove the clothing in the first place, in order to become the "savage beast." Such purposeful and systematic intent is obvious when he carefully hides them in the same place each time:

> "La est la piere cruose e lee,
> Suz un busson, dedenz cavee;
> Mes dras i met, suz le buissun,
> Tant que jeo revienc a meisun."

> ["Under a bush there is a big stone, hollowed out inside; I hide my clothes right there until I'm ready to come home."] (*Bisclavret* 93–6).

It is apparent, then, that living as a werewolf holds some attraction for Bisclavret, since he leaves his beloved wife[15] willingly to pursue the life of a wild beast for so much of his time. What activities, beyond living on "prey and abduction" mentioned in line 66 ("'Si vif de preie e de ravine'") attract him to that dark, hidden life one can only speculate: a gay lifestyle, bestiality, or sado-masochism are possibilities. The simple fact is that he chooses to divide his time between a life of full social integration, and one where he indulges in private pleasures of some kind. This wearing or not wearing clothes represents an exercise of his free agency, and as such is a major aspect of his existence, a primary sign of his two states of being.

Analysis of the depiction in the lai of the two natures shows that the life of the human Bisclavret is clearly based on appearances and social relationships. The first description of the social, human being tells of his beauty, what others see of him: "Beaus chevaliers e bons esteit" [He was a handsome knight and good] (author's translation) (*Bisclavret* 17). Next his noble behavior is noted: "E noblement se cunteneit" [. . . who behaved nobly] (*Bisclavret* 18). In the lai, "noble" refers to a social position in a system of hierarchical relationships, as does *chevaliers*, "knight," the noun Marie uses to represent him here. The next reference relates him to his pretty wife:

> Femme ot espuse mut vailant,
> E ki mut feseit beu semblant.

[He had an estimable wife, one of lovely appearance.] (*Bisclavret* 21–2).

Of course she could have been a wife who only appeared to be lovely, one who pretended well. The later indication of her dressing elegantly and bringing rich gifts for the king (*Bisclavret* 228, 230) might suggest a pretentiousness on her part as well. All of these meanings for *feseit beu semblant* nonetheless highlight the notion of social relationships and outward appearances. The final comment on the hero in human form describes his close relationship with the king and the esteem all his neighbors held for him:

> De sun seinur esteit privez,
> E de tuz ses veisins amez.

[He was close to his lord, and loved by all his neighbors.] (*Bisclavret* 19–20).

Bisclavret, the human hero, obviously maintains a thriving social network, vertically as well as laterally. Such an interweaving of lateral and vertical relationships, with his neighbors and with his wife and king,[16] already brings to mind the idea of weaving horizontal and vertical threads to form the fabric of the textiles which are so critical to the dynamics of this tale.

The beast *bisclavret* is described quite positively as well, however, but more in terms of inherent qualities, rather than appearance.[17] These descriptions, oddly enough, often comment on the characteristics of the wild animal as they are illustrated by his behavior in some social context as well.

> "Cum ceste beste s'humile!
> Elle ad sen d'hume, merci crie!"

["This beast is humbling itself to me! It has the mind of a man, it's begging me for mercy!"] (*Bisclavret* 153–4).

These words from the King not only attribute some natural mental acumen to the beast, but also point out that he is able somehow to perform appropriate social functions, such as reverencing or paying homage to him as liege lord. The King then mentions again the surprising degree of mental development he observes in the wild animal: "'Ceste beste ad entente e sen.'" (*Bisclavret* 157, This beast is rational—he has a mind.) The werewolf is next acknowledged as meriting the King's mercy: "A la beste durrai ma pes [I'll extend my peace to the creature]" (*Bisclavret* 159). While implying something of the werewolf's intrinsic merit, this pronouncement brings the wild beast into a social relationship with the King, not without some effort by the King, to bring him out of the dark privacy of the forest and integrate him into the public milieu.

Descriptive adjectives for the werewolf follow: "Tant esteit francs e deboneire [He was so noble (brave, good, nice, fine, author's translation) and well behaved] "(*Bisclavret* 179). Hanning and Ferrante use "noble" for the meaning of *francs,* perhaps to sustain an allusion to social interaction or ranking. In this context however the attributes listed don't necessarily contribute a great deal inherently to that allusion. The next lines propounding his model behavior and attitude to the King, however, do:

> Ensemble od lui tuz jurs alout:
> Bien s'aperceit que il l'amout.

> [He always accompanied the king. The king became very much aware that the creature loved him.] (*Bisclavret* 183–4).

Loyal service to and respectful veneration of one's superior are manifestations of sensitivity to what is appropriate for one's given position within a social complex or structure. And such a sensitivity would be more innate than for outward show.

A final characterization of the werewolf makes up part of Bisclavret's initial disclosure to his wife and raises the issue of the role of textiles in his double life. "'Dame, fet il, jeo vois tuz nuz' ['Wife,' he replied, 'I go stark naked']" (*Bisclavret* 70). In the absence of clothing he goes about as the wild beast. This statement contrasts with the references to the wife's pretentious dressing for the King. Here we see that the natural animal lives devoid of all pretense or concern for appearance whatsoever. However the notion of whether the being retains any residual personhood while in the form of a werewolf is called into question if we read "tuz nuz" as "completely no one."[18] Once again the issue presented is balanced even as it is introduced. The line ending with the suggestion of absence of personhood in the wild beast, *tuz nuz,* meaning "completely no one," begins with *Dame. Dame* can mean "wife," the vocative which affirms the social relationship between the werewolf and the woman. *Dame* is also the feminine title of nobility that establishes the addressee in the social hierarchy, which is nothing but a structuring of persons into set relationships. As the hero addresses his Lady and tells her he goes *tuz nuz* as the werewolf, he correlates the beast, that may or may not be without personhood, "completely no one," to the Lady, his wife, and inserts himself, as the wild animal, into the social structure. Moreover *dame, meaning par dieu,* "by God, for God's sake," would equate to Marie's refusal to condemn the interfacing of such a seemingly self-serving natural creature with the social entity. And textiles are even brought into that equation with the very next line, which invokes God directly by pronouncing the synonym of this last meaning of *dame:* "Di mei, *pur Deu,* u sunt voz dras? [Tell me, then, for God's sake, where your clothes are]" (*Bisclavret* 71). Why might Marie have included depictions of social roles, relationships, and functions in her accounts of Bisclavret as "bisclavret," the werewolf? It might have been precisely to broach the idea of a possibility or even necessity of integration of the one within the other, of the personal pursuit of selfhood and its desires within the restrictions or demands of social interaction.

The text contrasts a pair of words that focus at last on the textile itself. The terms Marie uses for Bisclavret's clothing are indeed consistent with the idea of a split: they are called only *dras* and *despoille,* or *despuille.*[19] Whereas the other lais exhibit quite a variety of terms, among them *dras* used quite extensively, only in *Bisclavret* do we find *despoille.* These terms create a connection between the social identity imposed from without, and the idea of a self-definition generated from within. In the Middle Ages clothing was usually given to identify the liegemen (and/or women) of a particular lord. The clothing was part of the social contract between lord and liege: the lord promised goods (arms, a horse, food and housing, if not actual lands with serfs to work them, in addition to the clothing) and the protection offered by his superior social status. The knight promised his services, to fight any of the lord's battles with and for him. As it is typically very much bound up in social obligations and duties to others, clothing "*vests*" the wearer with certain social rights and responsibilities

There are two uses of forms of the verb "vestir,"[20] but they are used more to highlight the contrast between the two natures or function of clothing. The word Marie uses to tie the vesting of social bonds to clothing is *dras.* "'Di mei, pur Deu, u sunt voz dras?' ['Tell me, for God's sake, where your clothes are']" (*Bisclavret* 70). *Dras* makes obvious reference to covering up, as in a fabric *draped* over the body, to cover it in some way. That function of covering up is associated with the social concern for appearance and even hiding or veiling the true or inner self. By extension, could such social ties, tagged by the giving and receiving of textiles, ever be anything more than superficial, never actually achieving any deep commitment involving the inner man or woman? The clothing the baron undoubtedly had at some time provided for his wife and/or that she had made for him apparently failed to establish any such exclusive or lasting bond.

In these same lines, *dras* is contrasted with undressing. However the verb, se despuille, "undresses," is identical to the spelling used for Bisclavret's clothing some 54 lines later. "Pur sa despuille l'enveia. [And then she sent the knight to get her husband's clothes.]" (*Bisclavret* 124). The idea of some form of clothing that somehow undresses, and reveals or expresses a true self, is brought out even more clearly by the more frequent spelling of the word, *despoille*. This spelling allows rich associations with skin, from *despouille* meaning "skin" or "hide," and with body hair, *poile* or, in the case of a werewolf, fur.

Calling the clothing *despoille* emphasizes not only *un*dressing or removal of clothing, from the prefix "des," but, in terms of subject construction, what the removal of the clothing strips down to: not merely to the skin, but to the bare, naked, true self, the subject, freed from imposed constraints or definitions, to act as a the self-directed agent. Even beyond this "soul skin," the removal of the clothing focuses our attention then on what grows out of that self in the place of clothing, the body hair or fur. Such a chain of thought makes the werewolf's fur a substitute form of clothing, one that comes from within him and could be viewed as expressing or representing his selfhood. By extension, the actions Bisclavret pursues in the undressed state or fur-covered form of a werewolf can be viewed as actions stemming very much from his own bare, naked, true self, from his own agency.

Even while trapped in the form of the werewolf, B/bisclavret is in full possession of his agency: his actions are never determined by his beastly form, which is to say, he never pursues the bloody pastimes natural to that form, at least when he is in some social setting.

> U ke li reis deüst errer,
> Il n'out cure de desevree;
> Ensemble od lui tuz jurs alout;
> Bien s'aperciet que il l'amout.

> [Regardless of where the king might go, He never wanted to be separated from him; He always accompanied the king. The king became very much aware that he loved him.] (*Biclavret* 181–4).

He obviously behaves with consistent loyalty and affection toward the King and, in fact, as affirmed in the following, to all:

> Tant esteit francs e deboneire;
> Unques ne volt a rien mesfeire.

[He was so noble and well behaved, he never wished to do anything wrong] (*Bisclavret* 178–9).

His savagery is manifest only by exception, toward his treacherous wife and her lover:

Unke mes humme ne tucha
Ne felunie ne mustra,
Fors a la dame qu'ici vei.
Par cele fei ke jeo vus dei,
Aukun curuz ad il vers li,
E vers sun seignur autresi.

[He's never touched anyone, or shown any wickedness, except to this woman. By the faith that I owe you, he has some grudge against her, and against her husband as well.] (*Bisclavret* 246–51).

It is clear that the werewolf's actions reflect his deliberate and conscientious adherence to the dictates of the rational and ethical inner "man."

A re-examination of some of these lines shows how *dras* highlights the superficial, social relationships, and *despoille* highlights the inner self.

Pur sa despuille l'enveia.
Issi fu Bisclavret trahiz
E par sa femme maubailiz.

[And then she sent the knight to get his clothes. So Bisclavret was betrayed, ruined by his own wife.] (*Bisclavret* 124–6).

Bisclavret's private, inner selfhood is violated when his clothing is taken because his agency to choose the form of his existence is overridden by the actions of his betrayers.

Coment ele l'aveit trahi
E sa despoille li toli.

[How she had betrayed him and taken away his clothes.] (*Bisclavret* 267–8).

With his clothing, his *despoille,* they take away his privilege to private selfhood. This deprivation occurs after his wife steals his public covering: "Puis que ses dras li ot toluz, . . . [And how after she had taken his clothes, . . .]" (*Bisclavret* 271). She has stolen his access to being a public figure in society, since he cannot return to human form and his human relationships without his clothing.

Bisclavret was close to his lord, ("De sun seinur esteit privez,") (*Bisclavret* 19), but the King values, even demands his most inner, naked, true self:

> Li reis demande sa despoille;
> U bel li seit u pas nel voille. . . .

> [The king demands his clothes; whether she wanted to or not. . . .] (*Bisclavret* 275–6).

By making the association between *voille* and *veil* (Greimas 670), and reassigning the subjunctive verb *seit* to B/bisclavret as subject (in both the grammatical and existential meanings of the word), to the stripped-down-to-his-inner-natural-self- werewolf, we read that, even in that form "whether he was beautiful or not, didn't matter, nothing was veiled" (author's translation). In other words, the King relates to Bisclavret in a very private fashion, even as a werewolf, where nothing is veiled and beauty doesn't matter.

The next revelation made by associating *dras* with outer, social relationships is Bisclavret's hesitancy to leave his "natural" state, even after being trapped in that form for a year or more:

> "Cist nel fereit pur nule rien,
> Que devant vus ses dras reveste."

> ["This beast wouldn't, under any circumstances, put on his clothes in front of you."] (*Bisclavret* 284–5).

Bisclavret is unwilling to take upon himself, to re-vest himself of, his public obligations by donning his public cover-up once again. There is good reason, however, for Bisclavret to ignore the clothing given to him publicly:

> Quant il l'urent devant lui mise,
> Ne s'en prist garde en nule guise. . . .
> Mut durement en ad grant hunte!

> [When they were put in front of him, he didn't even seem to notice them. . . . He's just too ashamed to do it here!] (*Bisclavret* 279–80, 288).

To dress in public would expose his nakedness as he shifted back into human form, which would further violate his privacy, his private inner selfhood. A parallel situation existed in Jewish law, where imposed nudity was punishment for adultery. To dress in public would be tantamount to admitting guilt for a social crime, of which he had been the victim, not the guilty party. It was his wife who was the adulteress. Hence Bisclavret refuses to expose himself in such a self-condemning action.

So the King, representing the outward demands of society, brings Bisclavret and the *despoille* into his private, inner chambers (already demonstrating a transformation of that society to accommodate the private being) for the long awaited transformation or possible final melding of his two natures. It is in this last scene where the private lifeform is perhaps publicly integrated at last. Conclusive observations as to that possibility await my final arguments, however.

Finally, Bisclavret is often referred to as a beast, *beste* in Old French. As shown above, Marie's descriptions of the *beste* are all positive. A very close reading of line 274 then brings together the man and the beast: "Que la beste Bisclavret seit. [That this beast was Bisclavret]" (*Bisclavret* 274). It is actually a subjunctive clause that can be understood to *will* that the social human, the capitalized Bisclavret, *should be* a beast, at least to some extent, to consider his own needs for self-fulfillment, and that this would in fact be *best.*" We should, of course, remember that Marie wrote for an English-speaking court, which would obviously understand such a double meaning, even though its official language was French.

Lines 285–6 offer another clause that could be read subjunctively:

Que devant vus ses dras reveste
Ne mut la semblance de beste.

[That he should put on his clothes in front of you in order to get rid of his animal form, to change his beastly form.] (author's translation) (*Bisclavret* 285–6).

In the subjunctive they would implore "that *before you* he should re-vest himself with his social obligations by draping a cloth covering over the form of the beast, but that this covering should not have to change the beast's character (an alternative meaning of *semblance*), or even "appearance." To comply with that subjunctive or imperative requires that "going public" with his hidden nature should not require risking the loss of self or of other love relations that Bisclavret has already alluded to:

"Mal m'en vendra si jol vus di,
Kar de m'amur vus partirai
E mei meïsmes en perdrai."

["Harm will come to me if I tell you about this, because I'd lose your love and even my very self.] (*Bisclavret* 54–6).

The risk is much more clearly stated, "If I tell you, I will lose my very self and leave you for loving myself" (author's translation). In order for him to resume his social position and responsibilities, without changing from his beastly appearance, and at the same time not risk losing either the love of fellow humans, or either aspect of his selfhood, means that, not only must the hero publicly acknowledge his dual nature, but that the public must than accept him in full awareness of it.

Lines 9–10 go further:

... ceo est beste salvage;
Tant cum il est

[This is a savage beast, such as he is.] (author's translation) (*Bisclavret* 9–10).

Relating the word for "savage," *salvage,* to the verb for saving, *salver* (with the suffix "-age" collecting the sum total of such saving actions or attributes), invites one to read a potential for some saving grace into the self-indulgent beast. An expanded translation of the above lines would read: "This is a beast who, even such as he is, as a sum total of his parts or aspects, is a model for saving the individual within the structure of social roles and responsibilities" (author's translation). To conclude: the beast is not merely savage; instead, attention to individual or personal needs, such as the beast manifests, is somehow saving, even in the context of social responsibility and awareness.

Line 29 shows the textiles as marker for all of the above:

Une feiz esteit repeiriez
A sa meisun, joius e liez. . . .

[One day when he returned home happy and delighted. . . . (*Bisclavret* 29–30).

Liez means delighted, but it can also be read as "tied down," which would refer, again, to his social obligations. Just as the cords that tie, the ties that bind, are spun textile elements, so the clothing he returns to marks his social role and responsibilities. However, it is the *tying together* of the two lifestyles that is joyful. It is the act of *returning* that is joyous and delightful, the ability and privilege to turn again and again, back and forth, from the needs of society to the private needs of the inner self. And such turning is also precisely the action of spinning thread, and of wrapping a person in cloth-ing.[21] Delighted, turned on, and by turns, tied down within the social bonds of duty: it is the freedom to come and go, "converse e veit" (*Bisclavret* 12), in and out of his textile ties, that delights him and brings him joy.

Conclusion

While wearing clothes, the hero is confined to his social role, husband and baron. When he chooses, he is free to roam the woods in the self-spun fur of a werewolf without his clothing. Although he may be trapped in a "savage beast's" body when it is stolen, it is his agency that still de-termines his actions. His preference to remain human for a slim major-ity of the time, four of seven days, indicates that he values the opportunities offered by human society slightly above those of the "self-indulgent" beast.

But in the semipublic acknowledgment of his duality in the King's chambers (rather than in his private revelation to his wife), Bisclavret is fi-nally saved from losing either aspect of his dual selfhood and/or his loving relationships. We may assume that he may thenceforth openly return, as in, turn and turn again, from one aspect of his being to the other, through the transforming power of his clothing.

It appears, then, that it is the careful balance, much like the necessary tension on the thread in spinning the ties that bind, in weaving the cloth for clothing, which offers an optimal life experience to Bisclavret. And it is only his free access to clothing that maintains his ability to orchestrate that delicate balance. Wearing the clothing, or not, articulates his choice in determining to what degree he will subject his individual desires to the order of society. Moreover, Marie's use of these textiles unveils or reveals, *despuille,* the urgency of such agency, unencumbered by public fears or condemnation, for optimum subject development.

Notes

1. Marie de France wrote somewhere in England and/or the north of France, presumably for the English court of Henry II (r. 1154–1189). We believe her to be a woman primarily on the basis of the feminine name she uses to introduce her work: "Oëz, seignurs, ke dit Marie [Listen, my lords, to what Marie says]" (author's translation) (*Guigemar* line 3). [Marie de France, *The Lais of Marie de France* (Durham: The Labyrinth Press, 1982)]. Sometime between 1160 and 1178 she composed 12 "lais," or short, perhaps originally sung "ditties" that each narrate a romantic tale. The Old French *Lais,* which, with her translation of King Alfred's English version of Aesop's Fables, comprise her major claim to fame, draw on

Celtic legends of heroism. Because of their extreme brevity, averaging 478 lines, the *Lais* invite close scrutiny of detail.

2. Elizabeth Wayland Barber, *Women's Work: The First 20,000 Years: Women, Cloth, and Society in Early Times* (New York: W. W. Norton & Company, 1994).

3. Actually, neither term fits our story's character, a being who must be naked to take on the animal form and then never speaks mute in that form.

4. The appropriateness of or logic behind such an action is born out by documents of a somewhat later period, which show such to be the established practice of Dutch courts, to cut off a woman's nose as legal punishment for adultery.

5. In " . . . senz nes sunt neies [were born without noses]" (*Bisclavret* 312–4) we may also read *nes* as "nothing at all," that they were born with nothing at all, without a heritage, because of their mother being "chaciee de la contree [chased out of the country]" (*Bisclavret* 306) for her crime against her husband.

6. The title, *Bisclavret,* seems to alternate randomly between a proper noun, the hero's name, which, in Jean Rychner's edition, begins with an upper case "B" and a common noun (accompanied all but once by the definite article, *li* or *le*) meaning "werewolf" in Breton, which Marie prefers to the Norman *Garwulf.* Although original manuscripts had no upper- or lower-case letters, the fact that Rychner chose to incorporate such a distinction in his universally accepted edition allows me to pursue how this recognized differentiation might further illuminate the nature of the man/beast.

7. We do find a contracted form of the definite article accompanying the proper noun here, "Del Bisclavret . . ." However it would seem only to underscore the fact that Marie wants to tell us the whole tightly tangled story, of the man-beast who cannot or perhaps should not be considered either/or. Hanning and Ferrante seem to grasp her introduction of this concept by refusing to clarify which being is the focus of the tale: they also include the article with the proper noun in their translation: "I want to tell about the Bisclavret."

8. "'The power elicited by the rite of homage is born out of the extremely intimate nature of the physical contact of the unequal participants. . . .The sense of touch, around which homage is centered, is the most sensitive form of personal communication" (Gloria Thomas Gilmore, "Conflicting Codes of Conduct: Marie de France's *Equitan*," *Utah Foreign Language Review* 2 (1990): 102.

9. I supply my own translation because Hanning and Ferrante do not follow Rychner's use of capitals in the Old French text. Perhaps in an effort to increase contemporary reader interest, they personify the beast by omitting the Old French article accompanying the common noun, and simply use "Bisclavret" as a proper name: " . . . where Bisclavret was" (*Bisclavret* 138).

10. See note 8 above, "they ran across Bisclavret" (*Bisclavret* 140).

11. See note 8 above. The translators also often then supply an additional term, which refers to the hero's beastly nature, in a following line: " . . . the knight whom Bisclavret attacked . . . the creature hated him."
12. See notes 8 and 11 above: "The king . . . went back to the forest where he had found Bisclavret, and the creature went with him" (*Bisclavret* 223–224).
13. See note 8 above: "She had them brought to Bisclavret" (*Bisclavret* 278).
14. In reading the one term for the distinction between the social human being and the wild beast, we find a perfect balance of emphasis: there are eight citations of each upper- and lower-case spellings. The tiebreaker would be line 75, where Bisclavret is capitalized, but perhaps only because it begins the line: "Bisclavret sereie a tuz jurs." It certainly would seem to refer to the beast, rather than the human hero of the story, because without the clothes he would not be able to return to the form of a man, to being Bisclavret: he would always be a *bisclavret*, a werewolf. Here Hanning and Ferrante refuse to break the tie: this time they override the capital letter found in Rychner's text and translate the line as "I'd stay a werewolf forever." Seemingly with the translators' concurrence then, I find that this line weaves back together the hair I have been splitting, to show that the character's two opposing natures will continue to exist in the one compound being, with emphasis on the need for balance between the two dimensions perpetually at play.
15. Despite later developments, she seems an ideal wife as the story begins:

> Femme ot espuse mut vailant
> E ki mut feseit beu semblant.
> Il amot li e ele lui.

[He had an estimable wife, one of lovely appearance; he loved her and she him.] (*Bisclavret* 21–3).

We may begin to suspect a facade however, for line 22 may also mean, "and one who pretended well" (author's translation).
16. The king is obviously superior to Bisclavret in the social hierarchy. His wife would have to be seen as his inferior in that same hierarchy, unfortunately.
17. It is oddly only the *garwalf*, not *bisclavret*, who is described at gory length as the evil maneater.

> Garvalf, ceo est beste salvage;
> Tant com il est en cele rage,
> Hummes devure, grant mal fait.

[A werewolf is a savage beast; while his fury is on him he eats men, does much harm.] (*Bisclavret* 9–11).

Restricting such negative behavior to the Norman term for the beast could of course be construed as some kind of political comment, on the bloody nature of the Norman character or their conquest of England just a century earlier. Marie's feelings regarding that event cannot be guessed, as we cannot know for a certainty which side of the channel claimed her allegiance.

18. *Nus* is the nominative case form for the indefinite personal pronoun, "no one." As an adjective *nu* even means *dépeuplé* (depopulated), which reinforces the idea of a removed personhood. Adding an "s" would only indicate that the adjective is being used as a noun in the nominative case.

19. There are other terms referring to being or getting dressed, such as *aturnez* or *s'appareilot,* used with the new husband and wife respectively, but *despuille* and *dras* are the only nouns used for Bisclavret's clothing (*Bisclavret* 193, 228).

20. Line 69 contrasts the idea of such social "vesting" with the undressed state of the werewolf, emphasizing the split in the two natures of Bisclavret. "S'il se despuille u vet vestuz [Whether he undressed or kept his clothes on.]" The lai closes with the restoration of the social clothing accompanied by their social vesting, in a proposal that their reintegration could not take place in full public view: "Cist nel fereit pur nule rien, Que devant vus ses dras reveste . . ." (*Bisclavret* 285–6). The semi-private reconciliation of the beast within the baron, when he dresses in the King's private chamber, is obviously an acceptable solution socially speaking, as the King rushes to kiss and embrace the reunited being:

Truevent dormant le chevalier.
Li reis le curut enbracier;
Plus de cent feiz l'acole e baise. (*Bisclavret* 299–301).

And it appears to be a peaceable solution for the sleeping knight as well.

21. Elizabeth Wayland Barber, *Women's Work: the First 20,000 Years,* 54, 132–3, 137–8. Barber points out that the first form of functional clothing was precisely that, a piece of fabric or even furs for wrapping around the body.

CHAPTER 5

FROM CONTENT TO FORM:
COURT CLOTHING IN MID-TWELFTH-CENTURY
NORTHERN FRENCH SCULPTURE

Janet Snyder

In the years between the 1130s and the 1160s, rows of column-figures appeared along the jambs of the doorways of churches in northern France.[1] These overlifesize painted limestone statues arranged as if in receiving lines may provide the best possible information about the appearance of courtiers at the time of Louis VII (r. 1131–1180). Rather than being represented wearing clothing copied from antique models, the column-figures appear to wear distinctive costumes of precious silks and finely-woven linens with embroidery or silk tapestry, as if elegantly garbed in contemporary courtly fashions. Just as in the twenty-first century one can distinguish the cowboy in chaps, Levi's and ten-gallon hat from the golfer wearing plus fours or the ambassador arriving from a fitting on Savile Row, during the twelfth century clothing could signal social position and power. The clothing of shepherds was distinct from that of landlords, and courtly matrons dressed differently from maidens. The examination of how and why clothing and textiles are represented in sculpture can be an effective tool in the search for the meaning of medieval portal sculpture.

The language of dress functions as a system of immanent signs, communicating subtle messages between contemporary "authors" and "readers." The authors of twelfth-century portal programs in France could count on the viability of this system of signs as they determined the appearance of column-figures. For the modern reader, the ideas communicated through textiles and dress can provide access to layers of meaning embedded in sculptural programs. Though the column-figures cannot be identified as portraits, much of the represented clothing illustrates the authentic ensembles worn by the elite of society. Unlike those preserved in the British Isles, inventories, wills, and wardrobe accounts for early twelfth-century French courtly households have been almost entirely lost, so few written comments exist that might document the value of textiles and clothing. The portal program itself constitutes an historical document, presenting the best information available concerning textiles and clothing.

While much has been lost due to war and environmental damage, today there is firm evidence for about 150 of the column-figures installed during the mid-twelfth century.[2] In the nineteenth century, the notion that these column-figures stood for the ancient kings of France was replaced by the theory that they represented personages of the Old Testament—the patriarchs, kings, queens, and high priests of the Old Law and the ancestors of Christ—in a sort of horizontal version of the Tree of Jesse.[3] It may be that the iconography was intentionally multivalent. Although Eugène Viollet-le-Duc looked on medieval images as genuine representations of daily life, the illustrations he published reveal his nineteenth-century sensibilities: his drawings resemble contemporary fashion plates.[4] During the 1890s, Wilhelm Vöge led the reaction against this naturalistic view of medieval sculpture with such force and eloquence that twentieth-century studies focused almost entirely upon the investigation of carving style, program iconography, and relation to liturgy.[5]

In this chapter, evidence corroborating the contemporary nature of the clothing represented in the stone column-figure sculpture will be provided. Parallels of these representations exist in various media. Contemporary clothing was illustrated in embroidered images, in painted images in manuscripts, on painted stained glass, and on Mosan enamel plaques representing known persons. There are proto-portraits on figured wax seals, and ivory chessmen show various the social roles. Similar personages appeared in related stone sculpture such as the archivolts at Chartres Cathedral, cloister column-figures, or the thirteenth-century statues on the inner west wall of Reims Cathedral. Works of nineteenth-century costume historians and linguists provide descriptive terminology. In addition, the *ordos* for the *sacre* of the French king and related written references in letters, sermons, and vernacular literature from the Middle Ages support the analysis.

In the middle of the twelfth century particular costumes define a distinct period of fashion in France. These costumes can be recognized in the sculpture of portals beginning with Saint-Denis and Chartres, and at cathedrals such as Bourges, Paris, or Angers. The costumes represented in these portal programs differ from reliefs made before 1130 at Moissac, and they differ from the clothing of prefigurations of Christ at Senlis, carved after 1165. The column-figures of the middle third of the century were carved with such precision that it is possible to observe the features of the garment construction and to identify some of the textiles.

In the twelfth century, the vocabulary of dress was a simple one. At the beginning of the century, the costume worn by both men and women of the upper echelons in northern French society comprised a *chemise*, a *bliaut*, and a mantle. The clothing of women is the most readily recognized marker of courtly dress among the column-figures. No two of the column-figure women are identical, but they can be sorted into three categories of variations on the simple gown: a one-piece close-fitting *bliaut*, a two-piece *bliaut* (*bliaut gironé*), and a looser tunic (*cote*). Each of these styles appears to signal rank and position of the women column-figures. The ensemble is illustrated in the clothing of the figure on the right in the photo of the sculpture on the right jamb of Saint-Maurice in Angers. (Figure 5.1.) The garment worn next to the skin, visible at Angers at her neckline and where she has lifted her dress at the hem, was the *chemise*, a long, roomy under-tunic with long narrow and rucked sleeves.[6] Worn over the chemise was the *bliaut*, an ankle-length tunic with a flaring skirt and long, hanging sleeves. The one-piece *bliaut* on the right jamb at Angers,[7] appears to have been made from a single length of cloth from neckline to hem, and it was cinched at the natural waistline. Along her left side, the lacing of the woman's *bliaut* is visible as it is nowhere else among the column-figures. Her mantle is a long cloak cut as a half-circle that is fastened around both shoulders with a double cord.

The representation of the women wearing the one-piece *bliaut* sets them apart; the majority of the women column-figures carved between the 1130s and the 1160s were depicted wearing a new form of the *bliaut*, the *bliaut gironé*, as seen on the figure on the right in the left portal of the west facade of Chartres Cathedral. (Figure 5.2.) The *bliaut gironé* was made in two pieces with a tight bodice (*cors*) and a skirt (*gironé*) that was finely pleated into a fitted, low waistband. The warp threads of the skirt fabric run horizontally; that is, the selvages (the finished sides of the piece of fabric) were used at the waist and as the *gironé* hem so the vertical pleats hung practically parallel to each other. The finely pleated woolen fabrics found at Gamla Lödöse and other West Scandinavian sites have been identified as the northern equivalent of the *gironés* worn by the women of the Chartres Cathedral west portal.[8] Most women also wore a double-wrapped girdle, (*ceinture*), with long pendant ends that were often tasseled or bound with decorative metal tips. The center of this girdle was placed over the woman's diaphragm in front; the girdle crossed around her back and returned to be knotted over her pelvic bone, emphasizing her womb beneath a softly rounded belly.

Figure 5.1   Column-figures: chevalier; man; woman in one-piece *bliaut*. Detail, right jamb, west portal, Saint-Maurice, Cathedral of Angers. Photograph by Janet Snyder.

Among the women depicted in these portal programs, a veil may completely cover a woman's hair, though several woman column-figures appear unveiled, which must be as significant as the two women who appear without the mantle.[9] The form of the veil distinguishes a young woman in a small square veil from a matron wearing a long rectangular veil draped across her chest beneath her chin, like the later wimple. Usually a woman's hair is divided into two tresses by a center part (grève).[10] The extreme length of braids depicted in column-figure sculpture may seem more believable as a fashion statement in the light of contemporary sermons urging women not to purchase hair (of dead women) with which to supplement their natural hair. Further, archaeological finds in London indicate tresses of silk, flax, or wool might be used to supplement human hair.[11]

During the twelfth century, wax seals were impressed upon charters by the highest-ranking members of society to confirm oral agreements.[12] After about 1130 these wax seals began to employ images of persons in what might be termed "proto-portraits," providing vital documentation for clothing during the Middle Ages. In the Sigillography Department at the Archives nationales, some of these seals are simply catalogued as "twelfth century," but most can be precisely dated between 1151 and 1190 because the documents to which they are appended have firm dates. Some of the seals show legends naming their owners, and the sigillographic descriptions identify women as *femme du roi, vicomtesse, comtesse, dame,* and *fille du roi.*[13] In their proto-portraits, these high-ranking "elite" women were represented wearing the *bliaut gironé* of the women column-figures of the Royal Portal at Chartres. (Figure 5.2.) The seal impressions were so precisely rendered that coiffures and the *bliaut* and mantle styles can be distinguished; the elite women's clothing clearly differs from that of lower-status women. The contrast among the seals between clothing worn by women belonging to different social strata signals a social significance comparable to the political significance expressed through the column-figures. The representation of persons wearing contemporary courtly clothing along the jambs of these church portals signals the conscious assertion of the precedence of the court (and with it, the Church) over the minor nobility.

The *bliaut gironé* is the characteristic courtly costume of the period, retained at court even after the fashion shift that had occurred around the marriages of the daughters of Louis VII.[14] The *bliaut gironé* is represented on the seals of Eleanor of Aquitaine, her daughter Marie de France, and two of the three seals of Agnès de Champagne. The seal of Ida, countess of Nevers, who died around 1178, shows a slim woman wearing a *bliaut gironé.*[15] The knot of her low *ceinture* is visible, with the pendant ends hanging in front. Her long, pointed, hanging *bliaut* sleeve cuffs flare in a

Figure 5.2 Column-figures: two ranking men; a woman in the *bliaut gironé,* left jamb of the left portal of the west facade of Notre-Dame, Cathedral of Chartres. Photograph by Janet Snyder.

bell-sleeve or trumpet shape, and very narrow *chemise* sleeves extend over the base of her hands at the wrist.

During the same time period, 1160–1190, sigillographic proto-portraits of the *chatellaines* of Cambrai show standing women wearing the simple tunic belted at the natural waistline.[16] The seals of Ada, wife of Simon d'Oisy, *Châtelain de Cambrai,* 1163, and women like her confirm the significance of the *bliaut gironé* as the costume of the elite. High-ranking women continued to use the mid-century court costume in their seals. This appears most consistently among the women in the immediate family of the monarch. Until nearly 1200, many of the sigillographic "proto-portraits" of close relatives of Louis VII continued to feature the conservative, formal fashion of the court, the same clothing represented in mid-twelfth-century sculpture.[17] Although in Germany, fashion seems to have shifted to the looser tunic belted at the natural waistline by the time *The Gospels of Henry the Lion* were illuminated in the 1170s,[18] Matilda of England was still shown wearing extremely long, bordered cuffs at her coronation. The rest of Matilda's gown is concealed by her mantle, but the other women at court follow her lead in wearing tunics with hanging cuffs.

Men also wore the ensemble comprising *chemise, bliaut,* and mantle. The *chemise* is visible on most of the forearms of the column-figures. The man's *bliaut,* worn on top of the *chemise,* usually fitted less tightly around the torso than the woman's. This simple *bliaut* was cut of a single length of fabric from neckline to hem. In order to enlarge the skirt of a garment, the tailor inserted triangular wedges of fabrics (gores). The resulting flared skirt had pleats that fell both vertically and in broad, smooth folds.

Georges Duby cited Gerard of Cambrai and Adelbero of Laon when he described the Orders into which medieval European society was organized.[19] According to this scheme, men were sorted according to three types of action: *agricolari-laboratore, orare,* and *pugnare:* those who work, those who pray, and those who fight. Quotidian workers in short gowns appear in the *June, July,* and *August* Labors of the Months at Chartres. A few column-figures like the trumeau depicting Saint Loup or the episcopal figure at Saint-Ours in Loches wear liturgical vestments that have parallels in enamel like the Mosan enamel plaque of Henri of Blois (now in the British Museum).[20] Most of the column-figure men wear the long gowns of high-ranking, elite men as seen on the pre-episcopal seals of Louis VII's younger brothers.[21]

Rank might be communicated through the manner of wearing the ensemble. The knee-length *bliaut* appears to signal rank within courtly circles, as it was illustrated on the Limoges enamel tomb effigy of Geoffroi Plantagenet, Count of Anjou,[22] and on members of the court attending the Coronation of Henry the Lion. The figures of Saints Peter and Paul at

Étampes and the two men on the left jamb of the left portal on the west façade at Chartres (Figure 5.2) are dressed in the ensemble of such close associates of the ruler, the shorter *bliaut* over the chemise. For most men, the mantle fastened on the right shoulder, freeing the right arm for writing, swordplay, or managing the reins of a horse.

In the twelfth century nobility as an institution was still being worked out, yet the elite, high-ranking men can be recognized in dress as well as in characteristic activities: among the archivolts at Chartres, the landlord pruning plants in the *April* Labor was represented wearing a skirt open to the hip, as was the Falconer of *May*. Initially, it might seem that men whose occupation was to fight were missing from the ranks of the column-figures, but closer examination reveals the *chevaliers*. The quintessential seal proto-portraits of elite men show a horseman in arms with the skirt panels of the long open *bliaut* worn beneath the chain-armor hauberk flowing on each side of the horse. Parallel images appear in the contemporary manuscript from Cîteaux illustrating hunting with falcons,[23] or as a knight among the ivory Lewis Chessmen at the British Museum: each warrior rides a horse. The language of dress records the horseless mounted warriors among the column-figures in shorthand.[24]

It is the open center seam on the man's *bliaut* that indicates a knight's garment: the skirt has seams open from foot to hip at the front and back center. These open center seams provide the man wearing a long gown with the freedom of movement required to straddle a horse. This functional aspect of the *bliaut* with open seams is less clear in a standing figure: when the sides of the fabric panels hang parallel, the seam appears to be closed. The edges of the fabric panels are easier to see when they are bordered or if the lower corners curl and separate. Bordered panels are illustrated in the clothing of the figure on the left in the photo of the sculpture on the right jamb of Saint-Maurice in Angers. (Figure 5.1.) At the Royal Portal of Chartres, in the clothing of the figure immediately to the right side of the center door, the lower corners of the fabric panels appear to curl and open slightly. Elsewhere, such panels appear to overlap. Since knights were required to maintain a number of retainers and several horses, their costume was an indicator of social rank and wealth: these are the elite of the elite who bore the double responsibilities of Christian knighthood. For the ideal Christian knight, the defensive characteristics of knighthood—war readiness, prowess, or belligerence—were balanced by the peace-loving aspects—intervention, use of reason, making and preserving peace. In this way, the right to own and bear arms brought with it the responsibility to care for the rest of society. During the twelfth century the king, that ideal warrior, was repeatedly charged with the care of the Church, widows and orphans, and the poor.[25]

During the twelfth century, men's hair varied in length from very short ringlets to coiling locks hanging to mid-back. Among the column-figures, hair style may serve as an identifying attribute. The only bearded man with a receding hairline on any twelfth-century portal is Saint Paul; across the doorway Peter wears a shorter beard, has curly, cropped hair, and holds a giant key. The kings among the Lewis Chessmen wear a hairstyle layered with short curls on top and great ropes of hair hanging in four long locks of separately coiled corkscrew curls over their shoulders.[26] A very few men column-figures have youthful, smooth chins. While the knights at Angers appear to wear very short-stubble beards; most column-figures sport individualized beards. These vary from a mass of curly knots through trim, combed beards divided into curls or corkscrews along the jaw line, to long beards that extend down over the chest in thick, snaky locks. Mustaches are generally in proportion to the beard, though sometimes men wear "handlebar" mustaches that extend to the jawline of a relatively short beard. Unlike Harold in the Bayeux Embroidery, among the column-figures, no unbearded man wears a mustache. Parallels in manuscripts, painted stained glass, smaller sculpted depictions, and the Bayeux Embroidery confirm the significance of hair and beard styles as indicators of male rank and station. Elsewhere in this anthology, Bonnie Effros discusses the legendary Frankish restriction of long hair to members of the royal dynasty as recorded by Gregory of Tours.[27]

Among the column-figures, the special status of an anointed king is designated through his ensemble, most easily recognized by his knee-length ceremonial dalmatic, and frequently by his hairstyle of four long, narrow curls resting on his shoulders. (Figure 5.3.) During the Middle Ages, the king of the French assumed a divine character through the act of his consecration.[28] In this reincarnation, he assumed an entirely changed aspect; the new clothes that he put on during the course of the *sacre* expressed the king's experience of a right of passage and revealed his altered state. The liturgical calendar that specifies the various parts of the coronation ceremonial and liturgy, the *ordo,* was revised several times during the Middle Ages.[29] As part of the rituals associated with his consecration, the king disrobed to wear only a white silk *chemise.* He was anointed with oil, and then he put on the royal costume over his *chemise*—a tunic, a dalmatic, and the royal mantle fastened by a brooch (*fermail*) on his right shoulder.[30] The ceremonial dalmatic was not mentioned by name until the coronation *ordos* of the fourteenth century, though it is described in the manuscript of the 1250 *ordo.*[31] The regal dalmatic represented the king's transformation as he became the anointed of God. Significantly, it appears as one of the garments worn by some of the column-figures.

Figure 5.3 Column-figures with regal dalmatic, right jamb of the Porte de Valois, north transept portal, the Royal Abbey of Saint-Denis. Photograph by Janet Snyder.

In twelfth-century sculpture, the regal dalmatic was represented as a narrow, but not form-fitting tunic shorter than the usual man's *bliaut*. It appears to have been constructed of practically flat, unpleated front and back panels of fine cloth attached at the shoulders, belted at the natural waistline, left open down the sides and finished with a heavy appliquéd band of decoration bordering the panels at the hem. In some examples, these dalmatics have rather short, relatively wide sleeves, much like those of a liturgical dalmatic, while in other examples dolman-cut sleeves terminate in narrow cuffs bunched at the wrist. The sculpted regal dalmatic and long hair of the column-figures of the Porte des Valois at the Royal Abbey of Saint-Denis (Figure 5.3) can be compared with the figure of a king in the Cloister of Notre-Dame-en-Vaux at Châlons-sur-Marne, or the painted representation of the seated King David from the *Stephen Harding Bible*.[32]

The costume of the column-figure king reflects those influences that shaped the French notion of the monarchy. First, the tunic was often the mounted warrior's *bliaut:* in Frankish tradition the sovereign was the elected chief of a clan of warriors. Second, over this *bliaut* was worn the dalmatic of the subdeacon: according to Church doctrine the anointed monarch functioned as a member of the clerical hierarchy. Third, the scepter and banderole or codex held by the column-figure king represent the impersonal "*res*" of the state and the written law of the Gallo-Romans. Because the observer might discriminate individual column-figures through sartorial cues, the various functions of personifications of the king were clear (his role as military leader, his sacerdotal role, or his role as judge).

Costumes and fabrics carried broad geographical and cultural implications for European audiences. Characteristics of textile folds—pointed linen creases, soft thick woolen rolls, tiny silk pleating, narrow parallel ridges and tucks, undersewn sleeve or mantle linings, flowing bias drapery of inset gores—are legible to the practiced eyes of modern textile historians in the column-figure sculpture. For the quotidian observer of twelfth-century sculpture not only the garments would have been recognizable, but the distinction between European cloth and goods from East of Venice could also be appreciated. The original polychromy of portal sculpture, now lost, would have accentuated the use of specialty goods.[33]

Courtly clothing in northern France during the twelfth century combined available goods with exotic imports: it was made of linen, wool, and silk, and it was decorated with metallic thread, pearls, and precious jewels. Before 1100, price and scarcity limited the use of silk and fine cloth to the highest-ranking and wealthiest persons in northern Europe. Silk, cotton, and cotton-linen blends came from the Middle East and the Orient; silk, linen, and *tiraz* came from Egypt: gauze originated in Gaza; damask in Damascus.[34]

Egypt had the reputation in the Middle Ages as the "land of linen," with the best grades produced along the Lower Nile.[35] Fine extant linens from Islamic *tiraz* correspond in weight and quality to textiles represented in twelfth-century French sculpture. By association, the fabrics and cut of their costumes identify the column-figures as personages from the East, the Holy Land. Concurrently the reciprocal inference associated the contemporary French elite with the prestige of these biblical models. The Old Testament warrior-leaders such as Abraham or Joshua were sometimes represented during the Middle Ages in military apparel—for example, in the tenth-century Byzantine *Joshua Roll,* the painted stained glass of the Sainte-Chapelle (ca. 1248), and the inner west wall of Reims Cathedral (ca. 1230).

The identification of the rank and standing of column-figure personages on the basis of costume represented in mid-twelfth-century French church portal sculpture reveals the complex networks of social, political, intellectual, and economic circumstances that frame these portal programs. It underscores the descriptive activity of the *ymagiers* who skillfully translated real clothing into stone. In turn, these monumental sculptures illuminate the way in which the Church, beginning with Saint-Denis and Chartres, used and shaped political reality in its imagery. Dressed in the clothing of the courts of northern France, most of these column-figures appear to represent the people closest to the ruler, and to set forth an enduring reminder that this earthly court is the image of the celestial court.

Notes

1. Angers, *Saint-Maurice;* Avalon, *Saint-Lazare;* Bourges, *Saint-Étienne, north portal;* Bourges, *Saint-Étienne, south portal;* Chartres, *Notre-Dame, west façade;* Corbeil, *Notre-Dame;* Dijon, *Saint-Benigne;* Étampes, *Notre-Dame-du-Fort;* Ivry-la-Bataille, *Notre-Dame;* Le Mans, *Saint-Julien;* Loches, *Notre-Dame du Château (Saint-Ours);* Nesle-la-Reposte, *abbey of Notre-Dame;* Paris, *Notre-Dame, the Saint Anne Portal;* Paris, *Saint-Germain-des-Prés;* Provins, *Saint-Ayoul;* Provins, *Saint-Thibaut;* Rochester, England, *Cathedral;* Saint-Denis, *Royal Abbey church, west façade;* Saint-Denis, *Royal Abbey church, Porte des Valois, the north transept portal;* Saint-Loup de Naud, *priory church;* Vermenton, *Notre-Dame.*

2. In addition to extant sculpture published in Willibald Sauerländer, *Gothic Sculpture in France, 1140–1270* (Paris: Flammarion, 1972), see portal illustrations in Dom Bernard de Montfaucon, *Les Monumens de Monarchie Françoise, qui comprennent l'Histoire de France* (Paris, 1729); and Dom Urbain Plancher, *Histoire générale et particulière de Bourgogne* (Dijon, 1739); and drawings of 1728 in Paris, BN, MS FR 15634, f 48–71.

3. In A. Katzenellenbogen, *The Sculptural Programs of Chartres Cathedral: Christ, Mary, Ecclesia* (Baltimore: The Johns Hopkins Press, 1959), Katzenellenbogen

reviewed the history of iconographic interpretations of Chartres and Saint Denis: In 1751, Abbé Lebeuf rejected Montfaucon's theory that the figures represented Merovingian kings and queens, suggesting instead that the statues on the royal portals should be regarded as personalities of the Old Testament; his opinion has prevailed for 250 years. For Mâle, Crosby, Kidson, and most recently Beaulieu, they are heroes of the Old Testament, the patriarchs and the kings, queens, and high priests of the Old Law, the ancestors of Christ. M. Aubert went on to identify each figure as each personifying a book of the Holy Scriptures. Abbé Bulteau, in line with Montfaucon, identified the crowned figures at Chartres as medieval rulers and queens. In an unsubstantiated comment, Kitzinger (Ernst Kitzinger, "The Mosaics of the Cappella Palatina in Palermo: An Essay on the Choice and Arrangement of Subjects," *Art Bulletin* XXXI [1949], 269–292) proposed that the biblical Kings at Saint-Denis are the antecedents both of Christ and the Kings of France. Katzenellenbogen, 27–8; fn. 2–3, 115.

Beaulieu's recent work continues in this exegetical vein as she names the column-figures. M. Beaulieu, "Essai sur l'iconographie des Statues-Colonnes de quelques portails du premier art gothique," *Bulletin Monumental* 142 (1984): 273–307.

4. E. Viollet-le-Duc, *Dictionnaire raisonné du mobilier français de l'époque carolingienne à la renaissance,* III-IV (Paris: Morel et Cie, 1872); The *bliaut,* 43; the *ceinture* 107; *Coiffure de dame noble, XII siècle,* hairstyle of a twelfth-century noblewoman, 188.

5. W. Vöge, *Die Anfänge des monumentalen Stiles im Mittelalter. Eine Untersuchung über die erste Blütezeit französisch Plastik* (Strassburg: Jeitz & Mündel, 1894).

6. *Rucking,* sewing or smocking on the inside of the fabric; that is, the fabric is gathered in a regular pattern on the inside of the fabric so the outside surface is regularly pleated or puckered. Terms used in French literature are "*ridé, froncé,* pleated." Eunice R. Goddard, *Women's Costume in French Texts of the Eleventh and Twelfth Centuries* (Baltimore: The Johns Hopkins Press; Paris: Les presses universitaires de France, 1927), 22. Quicherat uses the descriptive terms "plissé" and "gaufré par le fer de la repasseuse." Jules Quicherat, *Histoire du Costume en France Depuis les Temps les plus reculés jusqu'à la fin du XVIIIe siècle* (Paris: Librairie Hachette et Compagnie, 1875, 1877), 147, 163.

7. The one-piece close-fitting *bliaut* can be called a Western French or Plantagenet *bliaut* since it appears to characterize fashionable women's dress in areas under the influence of Henry II and Eleanor: in column-figure sculpture at the Cathedrals of Rochester in England and Angers in France, and represented on an enameled coffer said to come from Aquitaine now at the British Museum.

8. Ninety textile fragments were found at Gamla Lödöse, the precursor of Göteborg, situated on the Göta Älv River about 40 kilometers north of Göteborg. "The regular, parallel folds in these depictions have hitherto been regarded as primarily an artistic convention. But the discovery of

these contemporary, finely pleated fabrics shows that the artists were depicting a genuine style of costume." M. Nockert, "Medeltida dräkt i bild
och verklighet," *Den Ljusa Medeltiden* (Stockholm: The Museum of National Antiquities/ Stockholm, 1984), 191–6.

9. Only two women column-figures—one on the center left jamb at
Chartres Cathedral and the single figure remaining at Ivry (-la-
Bataille)—wear no mantle. Each of these women's costumes [both wear
the *bliaut gironé*] include other unusual features: At Chartres, the second
figure on the left jamb on the center portal wears neither veil nor *ceinture,*
and the Ivry woman wears her *ceinture* knotted at her natural waist.

10. These tresses normally emerged from below the hem of the veil at the
shoulders. Each of the tresses might be arranged *en trecié,* into long, heavy
braids arranged to fall along the outside profile of the arms, to the knees.
Braids *en trecié* used three sections of hair plaited without ribbons en-
twined. Simple *trecié* braids may have a ball or tassels pendant from the
bound-up tip of the braid. One way to avoid shortening the natural length
of hair with the folding and crimping of traditional braiding is to divide
the tresses again into two and to interlace the two with ribbons. Braids
made with the aid of a ribbon are said to be *galonné.* Goddard, 125. Very
commonly, the hair was dressed with lengths of hair not braided but hanging straight in bundles, like tubes, with ribbons delicately twined around
and into the hair, or ribbons of various widths wrapped in a criss-cross
fashion around the tubes of hair. In an even more complete deception,
women encased each tress in a tube of silk, which might be filled with artificial tresses and bound with ribbon to conceal the true length of a
woman's hair. See this style represented on the young woman ("the Bride
at Cana") column-figure from the cloister of Notre-Dame-en-Vaux at
Châlons-sur-Marne see Sylvia and Léon Pressouyre, *The Cloister of Notre-
Dame-en-Vaux at Châlons-sur-Marne, visitor's guide,* trans. Danielle V. Johnson
(Nancy: Mangin, 1981).

11. Plaited hairpiece with silk, tablet woven fillet no. 1450. "The use of false
hair was not a new departure in fourteenth-century England. Long plaits
worn down the back, sometimes almost to the ground in the twelfth century, often required the artful addition of extra hair, and from regulations
issued by a church in Florence in the early fourteenth century, it appears
that plaits of flax, wool, cotton or silk were sometimes substituted for hair."
Geoff Egan and Frances Pritchard, *Dress Accessories c. 1150- c.1450* (London: Her Majesty's Stationer's Office, 1991), 292–3.

12. Brigitte Bedos-Rezak, *Form and Order in Medieval France, Studies in Social
and Quantative Sigillography* (Aldershot, Hampshire, Great Britain and
Brookfield, Vermont: Variorum, 1993), 330.

13. Literally, Wife of the King, Vicountess, Countess, Lady, Daughter of the
King.

14. Louis VII had betrothed his four-year-old daughter Marie to Henri of
Champagne while in the Holy Land (1148), though this may not have

been officially sanctioned until 1153. In 1154, after settling a peace with Henry II, Louis VII made Thibaut of Blois his direct vassal, his *sénéchal*. It was probably at this time that Louis also arranged Thibaut's marriage with his second daughter Alix, who was then five years old. After her parents' divorce, Marie lived with a tutor in the convent of Avenay in Champagne until her marriage in 1164/1165. During these 11 or 12 years, Marie was recognized as the countess, Henri's wife, though the marriage was not consummated until 1164.

15. A. Coulin, *Inventaire de la Champagne recueillis dans les depôts d'Archives, musées et collections particulières des départements de la Marne, de l'Aube, de la Haute-Marne et des Ardennes.* Manuscript. No. 113 "Ida, comtesse de Nevers, femme de Guillaume III [1151–61, previous seal], morte au plus tôt en 1178. Dame debout, de face, robe ajustée à la taille, à manches pendantes, très longues; des deux bras écartés, la main droite tenant une fleur (PL XI). SIGILLUM IDE DE NIVERNIS COMITISSE. vers 1180." ["Ida, countess of Nevers, wife of William III, died ca. 1178; frontal view of woman in narrow, pleated gown, very long hanging sleeves, arms at right angles, right hand holding a flower, SEAL OF COUNTESS OF NEVERS, ca. 1180"] This is Demay n° 861.

16. G. Demay, *Inventaire des sceaux de Flandre recueillis dans les depôts d'archives, musées et collections particulières du département du Nord, ouvrage accompagné de trente planches photoglyptiques,* 2 vol. (Paris: Imprimerie nationale, 1873); Castellans of Cambrai: Flandre n°5502; "Oisy (Simon d') Châtelain de Cambrai, 1163. Sceau rond, de 68 mill. Arch du Nord; abbaye de Saint-Aubert. Type équestre; le cheval marchant à droit. SIGILLUM SIMONIS, CASTELLANI CAMERACENSIS. *Confirmation d'un don à Ham 1163.*" ["Simon d'Oisy, Chatellan of Camrai, 1163, round seal, equestrian type, horse traveling to the right. SEAL OF SIMON, CHATELLAN OF CAMBRAI. Confirmation of a gift to Ham in 1163."]

Flandre n°5503; "Ada, femme de Simon d'Oisy, Châtelain de Cambrai, 1165. Sceau ogival, de 67 mill. Arch du Nord; abbaye de Vaucelles. Dame debout, en robe et en manteau, coiffée en cheveux, un rameau à la main, sur champ de rinceaux. +SIGILLUM ADE, CAMERACENSIS CASTELLANE. *Exemption de droits de passage accordée à l'abbaye de Vaucelles.*" ["Ada, wife of Simon d'Oisy, Chatellan of Cambrai, 1165. Ogival seal. Woman facing front in gown and mantle, coiffed hair, branch in her hand, on a field of arabesques. SEAL OF ADA, CHATELAINE OF CAMBRAI. Exemption of rights of passage given to the abbey of Vaucelles."]

17. Including Eleanor of Aquitane. See Douët d'Arcq, *Inventaire de la collection des sceaux des Archives nationales,* 3 vol. (Paris: Plon, 1863–1868), n°10006 [Eleanor of Aquitaine, wife of Henry II] "Eléonore d'Aquitaine, (1199), femme de Henri II Plantagenet "(king 1154–1189). Seal and counter-seal. The seal is badly broken: only the middle of the torso remains in both examples. She is wearing a closely-fitting *bliaut* with a low waistline from which hang the vertical pleats of the skirt. Her arms are very slim, with

very long, narrow, hanging cuffs and a mantle that is fastened twice at the center of her chest (in one seal there is a diamond-shaped opening gapping between the fastenings). The figure is quite narrow and the textile is distinct and finely pleated. It may be that she used the same seal all her life; this reveals a conscious choice, for many women of lower rank revised their seals frequently.

18. See Henry's Coronation miniature in the *Gospels of Henry the Lion,* Herzog August Bibliotek, Wolfenbüttel, Germany, ca. 1173–5. Illustrated in Ronald W. Lightbown, *Mediaeval European Jewellery with a catalogue of the collection in the Victoria and Albert Museum* (London: The Victoria and Albert Museum, 1992), 104 and plate 11.

19. Georges Duby, *The Three Orders, Feudal Society Imagined, Tr. A. Goldhammer* (Chicago: University of Chicago Press, 1980), 4–5.

20. Neil Stratford, *Catalogue of the Medieval Enamels in the British Museum,* 2, *Northern Romanesque Enamel* (London, 1993), nos. 1, 2.

21. Douët d'Arcq, *Inventaire de la collection des ceaux des Archives nationales* (Paris: Plon, 1863–1868), Henri de France, c 1146, n° 7615; Philippe de France, 1137–1152, n°9181.

22. The tomb effigy of Geoffrey V [Plantagenet], Count of Anjou. Copper: engraved, chased, and gilt: *émail brun;* champlevé enamel: lapis and lavender blue, turquoise, dark and medium green, semi-translucent dark green, yellow, golden yellow, pinkish white, and white. Le Mans (?), shortly after 1151. Musée de Tessé, Le Mans (Inv. 23–1) Illustrated in *Enamels of Limoges, 1100–1350* (New York: Metropolitan Museum of Art, 1996), 98–9.

23. Dijon, Bibliothèque municipale, Ms 173, f.174.

24. My identification of a dozen column-figures as horsemen invites the reassessment of traditional interpretations of portal sculpture. For purposes of identification, the column-figures can be numbered from left to right, and labeled according to the viewer's left or right jamb when facing the portal. Men wearing a chevalier's costume appear: Angers Cathedral, Left 1,Right 1; Chartres Cathedral, west portal Center Left 3, CL 4, C Right 1, CR 2, RL 1, RL 2, RL 3: Bourges Cathedral south portal Right 1; Rochester Cathedral, Left; Saint-Bénigne de Dijon, Left 1.

25. For example, see Suger, *The Deeds of Louis the Fat,* trans. Richard Cusimano and John Moorhead (Washington, D.C.: The Catholic University of America Press, 1992), 154.

26. Mid-twelfth century, probably Scandinavian. British Museum Ivories 78–144. See the same hairstyle worn by, among others, Bourges South, left jamb, first figure, and the men column-figures at Rochester Cathedral and from Notre-Dame de Corbeil.

27. See Bonnie Effros, "Appearance and Ideology: Creating Distinctions between Clerics and Lay Persons in Early Medieval Gaul," in this anthology.

28. J. P. Bayard and P. de la Pierrière, *Les Rites Magiques de la Royaute* (Paris: Friant, 1982), 156–7.

29. E. A. R. Brown, *"Franks, Burgundians and Aquitanians" and the Royal Coronation Ceremony in France.* Transactions of the American Philosophical Society, fol. 82, part 7 (1992), 37.

30. The royal costume is described in P. E. Schramm, *Kaiser Könige und Päpste, Gesammelte Aufsätz zur Geschichte des Mittlealters* III (Stuttgart: Anton Hiersemann, 1969), 547–52. See also Janet Snyder, "The Regal Significance of the Dalmatic: the robes of *le sacre* represented in sculpture of northern mid-twelfth-century France." *Robes and Honor, The Medieval World of Investiture,* Stuart Gordon, ed. (Palgrave/St. Martin's Press, 2001), 291–304.

31. Paris, Bibliothèque national, MS lat. 1246. See Henri Comte de Paris, *Les Rois de France et le Sacre,* with Gaston Ducheta-Suchaux (Paris: Éditions du Rodier, 1996), 150–1.

32. *The Bible of Stephen Harding,* Bibliothèque municipale, Dijon, manuscript, 14.

33. "The costly, highly prized materials which were frequently imported from the Orient are often mentioned . . ." Goddard, 45. In the courts of Blois and Champagne, popular French literature included the *lais* of Marie de France, in which silk from Constantinople and Alexandria, fine linen and named garments are featured as key plot elements. See Marie de France, *The Lais of Marie de France,* trans. R. Hanning and J. Ferrante (Durham, N.C.: The Labyrinth Press, 1978), 7. See especially *Le Fresne, Lanval, Guigemar, Les Deus Amanz.*

34. Concerning the sources of textiles, see M. M. Postan and Edward Miller, eds., *The Cambridge Economic History of Europe,* 2nd edition (Cambridge, New York, etc: Cambridge University Press, 1987). See also Robert Sabatino Lopez, "Silk Industry in the Byzantine Empire," *Speculum* 20/1 (Jan. 1945): 1–42.

35. Made at Alexandria, Tinnis, Damietta and in Lower Egypt. See T. Thomas, *Textiles from Medieval Egypt AD 300–1300* (Pittsburgh: The Carnegie Museum of Natural History, 1990), 29. See also E. Sabbe, " L'importation des tissus orientaux en Europe occidentale du haut Moyen Age IX-X siècles," *Revue belge de philologie et d'histoire,* (juillet-déc, 1935), 1276. See M. Lombard, *Études d'économie médiévale, III, Les textiles dans le Monde Musulman du VIIe au XIIe siècle* Civilisations et Sociétés 61 (Paris: Mouton éditeur, 1978), 69–70. "According to the oft-quoted words of the Arab chronicler al-Tha'albi (d. A.D. 1037–1038): 'People knew that cotton belongs to Khurasan and linen to Egypt.'" Lisa Golombek and Veronika Gervers, *"Tiraz* Fabrics in the Royal Ontario Museum," *Studies in Textile History,* ed. Veronika Gervers (Toronto: Royal Ontario Museum, 1977), 83.

CHAPTER 6

FASHION IN FRENCH CRUSADE LITERATURE: DESIRING INFIDEL TEXTILES

Sarah-Grace Heller

F rom around 1190 to well into the fifteenth century,[1] audiences and readers in Northern France (and also even Occitania or Iberia[2]) with a mind
to reminisce about the Crusades could read or hear rhymed stories of the
sieges of Antioch and Jerusalem, and of the adventures of the men captured
at Civetot. Such narratives, part of what is now referred to as the Old French
Crusade Cycle,[3] mingle battle scenes and depictions of suffering with many
descriptions of rich armor and robes worn by Sultans, Amirs, and Frankish
knights, booty in the form of Saracen[4] silks and gold, and marvelously woven
and embroidered tents. For example, Frankish knights encounter the tent of
the Saracen leader Corbaran after they have saved his life:

> Lors fu ses trés tendus, paisonés et ficiés,
> Li pons d'or et li aigles par deseure dreciés.
> La tente fu molt rice, de bruns pailes ploiés,
> Et vers pailes ovrés desor l'erbe jetiés,
> A oisials et a bestes geronés et tailliés.
> Les cordes sont de soie dont il fu ataciés,
> Et la ceute porpointe d'un cier samit delgiés.

> [Then his tent was stretched, pitched and staked,
> at the top the cross-support of gold and the eagle were erected.
> The tent was very rich, draped with brilliant silk,
> and patterned green silk was thrown over the grass,
> with lengths of cut fabric worked with birds and beasts.
> The cords with which it was tied are of silk,
> and the quilt was sewn with a shining, delicate *samit*.] (*Chétifs*, 1707–13).[5]

What role did such descriptive passages play in the Frankish imagination, reliving the First Crusade even as new expeditions were constantly being planned? Is this propaganda, encouraging potential warriors with promises of textile booty? It is certainly a sign that audiences and poets alike enjoyed rich and specific images of textiles, clothing, and ornaments: this is popular pleasure literature, written in the vernacular, rhymed for successful performance, easily varied to suit a patron or changing tastes or vocabulary.[6] In this particular passage, several of the ten extant manuscripts vary the colors and the textiles described. Variants for "*de bruns*" (which can signify a type of rich dark fabric as well as "burnished" or "brilliant,") "*pailes ploiés,*" include "*de bons porpres roies*" in T, and "*de blans pailes roies*" in C. Scribes seem to have enjoyed imagining different colors for the tent, from something dark and draping or pleating ("*ploiés*"), to imperial Byzantine striped purple, to striped bright white silk. Can this be called a mentality of fashion?[7]

A number of scholars associate the beginnings of Western fashion with the exposure to the luxurious textiles of the Islamic cultures during the Crusades. Should the Crusades be given responsibility for ushering in a new mode of tastes and consumption? Extant textile evidence testifies that very significant quantities of luxury goods from the various Muslim merchant centers in Spain, North Africa, the Levant, and the Asian trade routes made their way into Europe all through the Middle Ages.[8] However, while the existence of trade may be proven by archeological and visual evidence, proof of the existence of a fashion system[9] requires inquiry into the attitudes and desires of the time, for which the best testimony is often written. Some of the fantasies surrounding the rich clothing of the "Saracen" lands, presented to twelfth- and thirteenth-century Northern French audiences through the relatively little-studied group of texts known as the Crusade Cycle, offer a fascinating and heretofore unanalyzed source for examining Frankish desire for rich Oriental apparel. The *Chanson d'Antioche* was purportedly originally the work of an eyewitness named Richard the Pilgrim, improved and expanded to its present form (including the *Chétifs* and the *Chanson de Jérusalem*) by Graindor de Douai around 1190–1200. If Western fashion is born with the Crusades (beginning in 1095–1100), these texts are ideal sources; they recount events from the period when fashion should have been born from about a century's distance, in a period when fashion should have become more fully established.

Merging the large bibliographies on Western and Islamic dress and textile history with those on chronicle-based Crusade history, moreover with a literary analysis of the Crusade Cycle (which up to now has only been treated descriptively or with a view to its historical veracity)[10] is admittedly an ambitious interdisciplinary proposition. The focus of this essay prevents treatment of these many sources for the entire period in question. Nevertheless, this study is an initial effort to cross these disciplinary boundaries, a crossing necessary to broaden understanding of fashion in the medieval period.

The case for dating the emergence of Western fashion to contact with Islamic culture at the time of the Crusades has been presented and debated by a number of scholars. Philippa Pullar presents the Crusades as an occasion when "rag-tag gypsies" could buy silk garments as only the rich had done before, leading to a marked increase in trade.[11] Authors Michael and Ariane Batterberry speak of the Crusaders' "almost instant capitulation" to Eastern luxury and finesse. They claim that Crusaders' wives rushed to "adopt native ways with alacrity," and that upon their return, a "craze for all things Arab swept across Europe."[12] Such views are problematic, the latter because it attributes a more important role to women in the early dissemination of fashion than seems to have been the case,[13] and further because both overgeneralize upon unclear sources.[14] Further, neither chronicle nor other literary sources depict the Crusaders as having time or opportunity to shop; this will be discussed below.

Nineteenth-century costume historians Quicherat and Enlart attributed the new, longer, trailing robes adopted by the Normans in the late eleventh century to their travels and conquests in Puglia and Sicily, a cosmopolitan area where Islamic peoples were a significant portion of the population.[15] Robert Bartlett seems to corroborate such an idea: he found that the princely courts on the frontiers of Latin Europe—in places such as Sicily, the Levant, and Spain—were centers of patronage, conspicuous consumption, cosmopolitanism, and fashion.[16] Max von Boehn saw the birth of fashion occurring when the Crusades took Westerners' natural "partiality for foreign clothes" and "natural desire for change" and brought it into interaction with the dress and manners of the East as well as many other countries, creating a new society with new requirements.[17] Françoise Piponnier and Perrine Mane similarly suggest that "the Crusades introduced the glamour and variety of textiles produced or used in the Islamic world to a large number of Western soldiers," but that it was really merchants who were responsible for bringing fabrics to Europe.[18] In both views, the Crusades represent a major impetus or inspiration to the creation of the Western fashion system, but in themselves did not provide enough material to sustain one; that had to be provided by an expanded system of trade.

Jennifer Harris only loosely associates the show of new ingenuity in tailoring with the period between the First and Third Crusades, never ascribing causality to the military pilgrimages to the Holy Land.[19] Zoé Oldenbourg surmises that displays of adaptation to Oriental luxury on the part of the Latin settlers seeking help from their European counterparts in later Crusades only served to antagonize the Franks of the West, rather than to win their admiration or inspire emulation.[20] François Boucher cautioned that much emphasis has been laid on the Franks' apparent bedazzlement upon encountering Eastern textiles. He pointed out that the thesis basing a "fashion revolution" on Crusade chronicles did not examine the historical context deeply: trade in Eastern goods preceded the Crusades. It is clear, nonetheless, that with the twelfth century the Mediterranean trade in foreign silks, spices, and other goods increased significantly, and contact with Byzantine and Arab cultures certainly encouraged Frankish taste for luxuries and rich textiles.[21]

What is the case for the Crusaders' apparent "bedazzlement" upon seeing the wealth of the Orient? In the Latin chronicles concerning the First Crusade, the main narrative source for most Crusade historians, "bedazzlement" largely translates to passages describing booty. But booty was essential to the campaign, as it was the only means at the armed pilgrims' disposal to replenish their supplies and avoid starvation, a threat that plagued their efforts and runs as a leitmotiv through the narratives.[22] The money raised to finance the project—through sales and mortgages, in such quantities that the value of goods fell—was often insufficient to get them to the Holy Land, let alone home.[23] Jonathan Riley-Smith states that the acquisition of booty "is recorded so often that one is tempted to believe that it played a very large part indeed in the Crusaders' thinking"; however, it is also important to note that "there is little evidence for them returning home with anything but relics."[24] In other words, even if the Crusaders were influenced by Islamic fashions enough to attempt to bring them back, once home they had little to show of what they had seen and most of what did come home went to churches.[25] Moreover, a relic is the antithesis of a fashion object: it is believed to be ancient rather than new, and designed to be conserved for all time rather than set aside when something new and better appears. This presents a serious challenge to any notion that the Crusades initiated the birth of Western fashion by introducing quantities of Arab textiles and clothing into the streets of Europe. The memory of booty, however, could very easily remain with them. Indeed, the narrative evidence testifies that remembering booty became a popular pastime. If the birth of Western fashion lies in the Crusades, we should look for evidence first of all in expressions of desire to consume Islamic products, rather than in the actual products. There is nothing inherently fashionable about Eastern silks or *tiraz* bands. They only become objects of fashion when they are desired, evaluated, worn, and admired, or eventually discarded in favor of something new. As Barthes said, "ce n'est pas l'objet, c'est le nom qui fait désirer . . ."—the name rather than the item is desired.[26]

One form of desire is covetousness, historically a source of conflict on the Crusades. This is not surprising, given that the idea of the campaign was a holy pilgrimage meant to be undertaken in the spirit of humility and piety, in direct opposition to the convention that warriors were reimbursed for their services with loot taken in battle. Examination of a scene in the *Chanson de Jérusalem,* where the Sultan attempts to trap the Franks by tempting them with riches, allows some insight into the place of desire for Islamic goods in the Frankish mindset as the poet tries to "correct" bad behavior by rewriting the events. Godfrey of Bouillon, the great superhero of the cycle, confronting the Sultan's wealth, makes the right choice and refuses the treasure; in the Spanish *Gran Conquista* the author gives a more realistic picture of the flagrant covetousness that occurred in the heat of the moment.[27]

Hoping to bring the Frankish leader under his sway and convert him to "Mahon" (Mohammed), the Sultan invites Godfrey of Bouillon to his court. Knowing that the Franks are naturally covetous,[28] his Turkish counselors advise him to bring out his white Arabian charger as bait:[29]

Couvert d'un rice paile de l'uevre de Cartage,
La sele l'amulaine u d'or a mainte ymage—
A esmals i sont fait oisel, poison marage.
Molt [est] li sele rice de bone oevre salvage—
Onques hom n'ala tant par terre ne par nage,
A camp ne en maison, par mer ne par boscage
C'onques veïst si bone, tant alast en voiage. . . .
N'i ot ne frain ne sele ne soit tot fait d'or mier,
Tot sont d'escheles d'or portendi li estrier,
Mainte esmeraude i ot et maint topase cier.
Li poitrals del ceval fist forment a prisier;
N'a si rice home en France quel peüst esligier,
Car venins ne pot home qui le port entoschier.
Plus fu blans li cevals que nois c'on voit negier
Et la teste avoit rouge con carbons en brasier.
D'un vermel siglaton ovré a eskekier
Fu covers li ceval, menu l'ont fait trencier:
Le blanc par mi le roge veïssiés blançoier.
Li frains qu'il ot el cief vaut l'onor de Pevier:
Poi est d'omes el mont nel deüst covoitier.

[(The horse) was covered with a rich silk of Carthaginian make;
the Saracen governor had a saddle of gold covered with many images—
birds and maritime fish are worked on it in enamel.
The saddle is very rich and of very fine foreign manufacture—
Never did a man, for all that he traveled by land or by boat,

in camp or at home, by sea or by forest
ever see one so fine, no matter how many voyages he had taken. . . .
There was never either reins or a saddle made of better gold,
it was all done in scales of gold hung all over the outside,
there were many emeralds and many shining topazes.
The horse's chest harness was extremely admirable;
there was not a man in France rich enough to have bought it,
for venom cannot poison the one who uses it.
The horse was whiter than snow that you see falling
and its head was red as coals in a furnace.
With a checkered vermillion siglaton silk
the horse was covered, they had cut it very well:
you could see the white shining out between the red.
The bridle it had on its head was worth the honor of Pithiviers:
Few men in the world would not covet it.] (*Jérusalem*, 7187–93, 7205–17)

There is much to observe from this passage. First, its length, which is representative of many passages describing clothing and personal adornments
in the cycle. Such length suggests clearly that audiences and readers appreciated this kind of amplification. Also noteworthy is the amount of description devoted to the horse's accessories, in proportion to that afforded
the actual horse. The medieval war charger was hardly visible under its silks
and gold harnesses: a challenge to describing any of its other aspects, perhaps. Nonetheless, in other times and places readers interested in a horse's
value would want to know about its gait, its bloodlines, its height and
weight, and so on. What interests here is color and texture, and little else.
The writers and readers here had eyes only for appearance, and that appearance was largely the result of human artifice rather than of nature or
animal husbandry. This passage could qualify as what some of the above-
mentioned scholars called "bedazzlement" before Islamic textiles: the narrator makes clear that these are superior in every way to anything that
could be found in Europe.[30] The fantastic note about the harness' preven-
tive medicinal qualities adds to the dazzling impression.[31] This passage, and
others like it in the cycle, are noteworthy for their specificity.

   Amplificatio, the Latin rhetorical term for this kind of literary embroidery,[32] is dismissed by some thinkers as mere fluff, or as formulaic and
therefore devoid of significance. There are indeed some formulaic passages
in the cycle, usually descriptions of a series of knights as each is dressed,
armed, and charges into battle.[33] But this contains more than the typical
burnished green helmet, well-laced hauberk, and silk standard. The poet
imparts information such as the provenance of the textiles (Carthage), the
materials and stones involved (gold scales, emeralds, topaz), specific technical vocabulary (siglaton, esmals), and the effect of light on the textiles, as

well as specific figural motifs (birds, fish). There is no other description quite like this in the cycle, or elsewhere. This shows concern for originality both on the poetic level and on the cultural level. It shows a poet who was capable of imagining and describing unique appearance, and an audience that admired uniqueness. This kind of sensitivity to novelty is an important indication of a fashion system, and is particularly emphasized by the elaboration of the novelty topos ("Never could a man find such a one . . ."). The concern for specificity suggests a process for evaluating the items (Carthaginian work sells for so much, siglaton is worth more than . . . and so on). This is made explicit when the narrator says that no French man could ever afford it. This suggests that the vernacular public of this time had some consciousness of shopping, which is to say of calculating values in order to make personal purchases, an important clue for the presence of a fashion system because it is part of the act of making a personal choice to reflect individual taste, as opposed to an honorific vestimentary system, for example, where clothing is distributed by authorities to demonstrate status and favor in a hierarchy, or in a system where choice is simply not an option.[34] The fact that this display of textiles, accessories, and horseflesh is effective in eliciting covetousness in the Franks—furthermore that the Turks are depicted as viewing the Franks as naturally covetous—suggests that the Northern French vernacular public of around 1200 viewed itself as a culture of people seeking new and original items of personal display. They were conscious of being part of what was likely a nascent fashion system. This description does not, however, present the horse and its trappings as items reflecting the personal taste and personality of the Sultan. They express his wealth and power. In coveting the horse and its silks, however, each covetous individual can imagine himself possessing something that makes him look like he has the taste of a sultan, with the aura of wealth and power that would accompany such an appearance. In the circumstances of the Crusades, killing powerful-looking men and looting their possessions presented a whole new kind of fodder for the imagination.

This brings up a question: if the Crusaders indeed learned fashion in the Orient, does it necessarily follow that there was an Oriental fashion system? What sort of vestimentary system was in place in the lands that the Crusaders encountered? Recent scholars have made some important studies of the honorific vestimentary systems that characterized the medieval Islamic empire.[35] Honorific garments, called *khil'a* in the singular, were bestowed regularly and in such quantity that certain caliphs kept their own textile factories. Fabric quality, fiber content, color, and the decorative woven *tiraz* bands constituted a visual semiotics of rank and importance at court. The system seems to have similar features, from Chinese nomads of the second century B.C.E. to Sasanian Iran/Persia to Byzantium to the lands of the Islamic empire and up to the early modern period. In Western Europe, robing existed intermittently but did not follow the same rules, according to Stewart Gordon. While it flourished in the church, its secular use was complex and shifting, but probably was coherent and meaningful to the public.[36] An honorific vestimentary system does bear certain resemblances to a fashion system. Both are based on a theatrical logic of display, require an audience to appreciate the codes involved, and use fine adornment to gain attention or seduce; both encourage production and comsumption and foster artisanal creativity in the culture; both develop complex semantics of details. However, in a fashion system personal choice is cultivated and valued, while in an honorific system the individual graciously accepts the choice of an authority. In a fashion system, value lies in novelty and the old is constantly discarded. In an honorific system, value lies in tradition: *khil'a* literally signifies something taken off one person and put on another.[37] Outside the courts, where choice was not imposed by a vertical hierarchy to the same degree, and in certain cosmopolitan centers such as those of the 'Abbasids in Baghdad, fashion systems seem to have existed from time to time, although not with the continuous, uninterrupted growth that the Western fashion system has shown.[38] Yedida Kalfon Stillman has observed that while fashion flourished in the 'Abbasid-dominated ninth and tenth centuries, political turmoil due to the rise of the Turkish military elites and the aggression of the Crusades in the late eleventh and twelfth centuries brought about greater general rigidification as well as enforcement of humiliating dress codes for non-Muslims.[39] If fashion systems can be said to exist periodically,[40] and if some did in the Islamic empire, at the time that the Crusaders arrived they were not functioning freely. If the Crusaders "learned fashion" on their pilgrimage, they must have learned it by observing an honorific system and its accompanying rich textiles, and ascribing their own meanings on to its semantic objects. The East provides objects, a point of inspiration, but not the method. The Franks see their covetousness, their desire for display and novelty, as something all their own.

The honorific system does become a part of the remembered experience of the Crusades as worked out through the cycle. Representations of gift exchanges occur in a number of places,[41] but the frequency is particularly remarkable in *Les Chétifs*.[42] Whereas the songs of Antioch and Jerusalem are primarily concerned with the movements of the whole army or of the famous leaders and thus offer little in the way of personal encounters between Easterners and Westerners, *Chétifs* imagines intimate friendships between a group of lost Crusaders and sympathetic Saracens. It recounts the story of a group of prisoners who are freed when one of their number represents the disgraced emir Corbaran of Oliferne[43] in judicial combat and bests two Turk champions, proving Corbaran's claim of the ferocity of the Christians. A number of adventures follow and the now friends repeatedly save one another's lives. The first episode of gift-giving between Corbaran and his champion, Richard of Calmont, is a representative example. When Richard agrees to do combat, Corbaran kisses him and dresses him in his own clothes ("Corbarans d'Oliferne son mantel desfubla,/ Par les resnes de soie al col li ataca, [Corbaran of Oliferne took off his mantle and tied it around (Richard's) neck with the cords of silk,"] [558–9]—a ceremony in the manner of that of the *khil'a*. Richard, however, keeps taking the mantles off and giving them to his other companions, saying that he will not accept even "vair ne gris" fur until everyone is dressed (558–68). He obviously plays by the rules of Christian humility and comradeship rather than reserving the honor to himself. Corbaran orders all the freed prisoners to be dressed in "pailes de bofus" (570, 599, a silk). In the end, Richard does allow himself to be dressed: "Ricars ot un bliaut trestot a or cousu,/ Li mantels de son col et la pene qu'i fu/ Et li tasel a brasme, ki sont a or batu,/ Valurent bien. M. livres de fin argent fondu [Richard got a *bliaut* sewn all over with gold thread, the mantle around his neck and the cloth that was there and the jeweled border, which are of beaten gold, were worth at least a thousand pounds of fine silver bullion"] (603–7).[44] The description suggests *tiraz* bands, the most valuable of which contained gold thread and were often used as borders on cloaks.[45] The passage is also another example of the cost-consciousness discussed above. The gift is interpreted by the narrator more in terms of its monetary than its honorific value, more in terms of its spectacular and luxurious appearance than its value as a signifier of hierarchy or something similar. The Frankish eye interprets it according to its own cultural logic.

Like Godfrey, Richard struggles between the Christian ideology of renouncing worldly goods and an appreciation of luxurious adornment. The text works the problem out nicely: renounce vanity long enough to get fine robes for everyone. In *Les Chétifs* the foreign convention of gift-giving becomes a site of fantasy, where common knights can dream of receiving fine Oriental robes, not only the greatest heroes. There is certainly a convention of giving gifts to companions in Europe by the later thirteenth century;[46] although it cannot be said whether this was a result of contact with Eastern honorific vestimentary systems, or whether it occurred independently. Regardless of any semblance of historical realism reflected in passages amplifying acts of gift-giving, they nonetheless advertise the notion that befriending a Saracen who is longing to convert from worship of the inferior gods Mohammed, Tervagant, and Apollin[47] is a way to earn fabulous treasures. The Tafurs are also used in the cycle to play with this struggle.[48] A marginal section of the army, also known as the "ribalds," they renounce worldly goods, fight with ignoble knives and axes, go barefoot, wear rags (*Jérusalem*, 1815–18, 1830–6), and gleefully eat dead Saracens in times of famine. Their generally vile nature—unwashed hair and all—is celebrated almost like a secret weapon in the texts. On two occasions, when the Christians are desperately threatened, Godfrey, now king of Jerusalem, dresses the Tafurs (who are among the few hundred men to stay in the city when the major armies head for home) in looted clothing and parades them around the city ten times, bluffing greater manpower than they have (*Jérusalem*, 6378–6411, 7290–7300). It is emphasized that he asks the counsel of *all* his men, "Chevaliers et ribals—petit i ot garçons (6377)," knights as well as the most lowly and marginal men. In this representation, the Crusade actively becomes a locus where men of all stations might situate dreams of wearing ermine or siglaton—if only in jest, to throw it down later and resume more righteous rags (6410–11). This follows the principle that a fashion system is a great social equalizer, allowing anyone with access to clothing to wear it and in the process undermining all hierarchical codes of appearance.

The evidence in the Crusade Cycle suggests that some elements of a fashion system were at work in the Northern French mindset of around 1200, and that these elements were being linked in fantasy to romanticized retellings of the First Crusade. Passages demonstrate desire for the new, or at least the exotic; this desire is open to men of all stations. A sense of conspicuous consumption is in place. There is evidence of desire for uniqueness and originality. Some aspects of a fashion system cannot be discerned from these texts alone, however. There is little evidence of the possibility for personal choice and expression of taste, or of emotional connection to adornment. These texts, in any case, only suggest that a fashion system mentality was connected to Crusade narratives a century after the actual encounters occurred; they are not a sure source for indicating what experiences of early "fashion" the first Crusaders might have undergone. In short, medieval authors and audiences may support modern costume historians in *associating* fashion with the Crusades; but attributing *causality* for the beginnings of Western fashion to the Crusades is problematic. The Crusaders, in their very eagerness to undertake the expedition into unknown lands, were embracing one particular novelty in great numbers, as Marcus Bull has suggested and as contemporaries themselves observed. Bull notes that "the perceived novelty of the First Crusade is all the more remarkable because people in western Europe in the central middle ages were seldom comfortable with innovation for its own sake."[49] Something happens in European mentalities around the time of the First Crusade, in any case, and it is marked by a new receptivity to novelty as well as to consumption. From that mentality, a fashion system would develop in Europe in the subsequent centuries, this is sure enough. But it would seem that this receptivity to novelty was already present when the pilgrims undertook their journey, rather than being something they discovered along the way. Europeans brought their own fashion system to the encounter with Islamic textiles; the Islamic textiles were not of themselves responsible for the European system. Fashion lies in the Franks' desire, more than in the dazzling objects that inspired their desire.

Notes

1. The Old French Crusade Cycle contains many branches. I discuss the works of the central nucleus here, referring to the editions of Suzanne Duparc-Quioc, *La Chanson d'Antioche. Edition du texte d'après la version ancienne. Documents rélatifs à l'histoire des croisades* (Paris: Geuthner, 1976), abbreviated "*Antioche*"; Geoffrey M. Myers, *The Old French Crusade Cycle, 5: Les Chétifs* (Tuscaloosa and London: University of Alabama Press, 1981), abbreviated "*Chétifs*"; and Nigel R. Thorp, *The Old French Crusade Cycle, 6: La Chanson de Jérusalem* (Tuscaloosa and London: University of Alabama Press, 1992), abbreviated "*Jérusalem.*" Translations are my own. On the dating of the cycle and the manuscript tradition, see Emanuel J. Mickel, Jr., Jan A. Nelson, and Geoffrey M. Myers, eds., *The Old French Crusade Cycle, vol. 1: La Naissance du Chevalier au Cygne* (Tuscaloosa and London: University of Alabama Press, 1977), xiii–lxxxviii.

2. There are surviving fragments of an Occitan *Song of Antioch* (Paul Meyer, "Fragment d'une chanson d'Antioche en provençal," *Archives de la Société de l'Orient Latin* 2 (1884): 467–509); given the generally poor survival rate of Occitan manuscripts, the existence of fragments suggests that more copies were likely in circulation at one time. The *Gran Conquista de Ultramar* is a Spanish rewriting of the entire cycle; Louis Cooper and Franklin M. Waltman, eds., *La Gran Conquista de Ultramar, Biblioteca Nacional MS 1187* (Madison: The Hispanic Seminary of Medieval Studies, 1989). On both versions, see Suzanne Duparc-Quioc, *Le cycle de la croisade* (Paris: Champion, 1955), 171–205 and 45–55, 84–5, respectively.

3. Léon Gautier was the first to call the collection of works "le Cycle de la Croisade," *Les Épopées françaises. Etude sur les origines et l'histoire de la litterature nationale,* 2nd ed., 4 vols. (Paris: 1878–1882), 1: 121–8.

4. I use the term "Saracen" to indicate the Frankish representation of the Muslim peoples encountered in the Levant, as it is used in Old French. It is often used synonymously with "pagan," "Turk," and "Persian," less often with "Arab." On the origins and significance of the term, Paul Bancourt, *Les Musulmans des les chansons de geste du cycle du roi,* 2 vols. (Aix-en-Provence: Université de Provence, 1982), 1: 1–32; William Wistar Comfort, "The Literary Role of the Saracens in the French Epic," *PMLA* 55 (1940):

628–59; Mark Skidmore, *The Moral Traits of Christian and Saracen as Portrayed by the Chansons de Geste* (Colorado Springs: Dentan, 1935), 26.

5. See other descriptions of rich tents hung with textiles, *Antioche* 4866–78, where the sultan's tent was hung with green checked silk ("de vert, ovré a esciekier"), yellow ("gaune"), blue ("inde"), and white, and could shelter 20,000 Turks; *Jérusalem* 6085–6172, a lengthy description ranging over four *laisses*, in which the sultan's tent is of bright silk worked in gold and set with light-emitting stones.

6. There are at least 16 known extant manuscripts, fragments, or mentions of the cycle in medieval library catalogues, quite a significant number compared to many *chansons de geste*. Charles V's library contained no less than 14 different romances, histories, and chronicles related to Godfrey of Bouillon and the First Crusade, many of which do not correspond to anything extant (Myers, *The Old French Crusade Cycle, vol. 1*, xiii–lxxxviii, esp. lx). *Chétifs* in particular shows considerable variability amongst its ten extant versions (Myers, introduction to *Chétifs*, xv–xxii); given that it is less based on historical events than *Antioche* or *Jérusalem*, it seems to have been more open to manipulation by poetic imagination. For a good discussion of medieval reading practices and aurality, see Joyce Coleman, *Public Reading and the Reading Public in Late Medieval England and France* (Cambridge, Eng.: Cambridge University Press, 1996); Paul Zumthor, *Essai de poétique médiéval* (Paris: Seuil, 1972), 37–8, 70; also the essays in *New Literary History* 16:1 (1984).

7. Concerning fashion and the "fashion system," see discussion in note 9.

8. There is a large body of work on medieval Islamic textiles, much based on various museum collections. For recent bibliography, Patricia L. Baker, *Islamic Textiles* (London: British Museum Press, 1995); also Maurice Lombard, *Etudes d'économie médiévale: Les textiles dans le monde musulman du VIIe au XIIe siècle* (Paris: Mouton, 1978); for a geographical treatment, R. B. Serjeant, *Islamic Textiles: Material for a History up to the Mongol Conquest* (Beirut: Librairie du Liban, 1972).

9. By "fashion system," I am distinguishing the tendency to seek adornment—probably present to some degree in most societies—from a social system that revolves around the consumption and production of novelty, the expression of self through display of originality, and so on. For a full development of this notion, see Sarah-Grace Heller, "Robing Romance: Fashion and Literature in Thirteenth-Century France and Occitania" (Ph.D. Dissertation, University of Minnesota, 2000). See also Roland Barthes, *Système de la mode* (Paris: Seuil, 1967); Gilles Lipovetsky, *L'empire de l'éphémère* (Paris: Gallimard, 1987); Herbert Blumer, "Fashion: From Class Differentiation to Collective Selection," *Sociological Quarterly* 10 (1969): 275–91.

10. The Crusade Cycle is still in the process of being edited, published by the University of Alabama Press. For the general plan of the work, see Emanuel J. Mickel, Jr., Jan A. Nelson, and Geoffrey M. Myers, eds., *The Old French*

Crusade Cycle, vol. 1, xiii–lxxxviii. For summaries, Karl-Heinz Bender, "La Geste d'Outremer ou les épopées françaises des croisades," in *La Croisade: réalités et fictions. Actes du colloque d'Amiens, 18–22 mars 1987*, ed. Danielle Buschinger (Göppingen: Kümmerle, 1989), 19–30; Alfred Foulet, "The Epic Cycle of the Crusades," in *A History of the Crusades*, ed. Kenneth M. Setton, Harry W. Hazard, and Norman P. Zacour (Madison: University of Wisconsin Press, 1989), 98–115. For studies on the work's historical authenticity and style, Suzanne Duparc-Quioc, Le cycle de la croisade (Paris: Champion, 1955) and *La Chanson d'Antioche: Étude critique*, 2 vols. (Paris: Geuthner, 1978).

11. Philippa Pullar, *Consuming Passions: A History of English Food and Appetite* (London: Hamish Hamilton, 1970), 82.

12. Michael Batterberry and Ariane Batterberry, *Fashion: The Mirror of History*, 2nd ed. (New York: Greenwich House, 1982), 82–5.

13. Recently, scholars of medieval dress have come to agree that men were more involved in the fashion system in this period than were women. Odile Blanc, *Parades et parures: l'invention du corps de mode à la fin du Moyen Age* (Paris: NRF, 1997), 31, 34, 191–2; Perrine Mane and Françoise Piponnier, *Dress in the Middle Ages*, trans. Caroline Beamish (New Haven: Yale Univ. Press, 1997), 77–79; also Sarah-Grace Heller, "Light as Glamour: The Luminescent Ideal of Beauty in the *Roman de la Rose*," *Speculum* 76 (2001), 951–2, and "Fashioning a Woman: The Vernacular Pygmalion in the *Roman de la Rose*," *Medievalia et humanistica* 27 (new ser.) (2000): 1–18. This view is supported by the Crusade Cycle: whereas male (even equine) dress is regularly described at length, women's dress is mentioned very infrequently, although women are regularly present in battle and in towns. In *Chétifs*, 3308–10: the women of Damascus are described as wearing fur mantles; beautiful pagan women are seen coming to town "en drap de soie estroitement vestue" in *Jérusalem*, 2935–7, and the Christian women are mentioned wearing wimples and carrying stones in their sleeves, *Antioche* 8306–7.

14. Batterberry and Batterberry are not specific about their sources. I know of few examples in the chronicle literature of Crusade wives instantly adopting Eastern ways; in one Arab chronicle, the author recounts a story told him by a bathkeeper of a Frankish knight who decides to imitate the natives in the bath and have his pubes shaved, and then asks that his wife be likewise shaved. There is no indication in the anecdote whether the wife's attitude was one of "alacrity," however. (Usamah ibn Munqidh, *Memoirs of an Arab-Syrian Gentleman, or An Arab Knight of the Crusades*. trans. Philip K. Hitti (Ithaca: Columbia University Press, 1927; rpt, Khayats: Beirut, 1964), 165–6).

15. Jules Quicherat, *Histoire de costume en France* (Paris: Hachette, 1877), 147; Camille Enlart, *Manuel d'archéologie française depuis les temps mérovingiens jusqu'à la Renaissance: vol. III, le costume* (Paris: Auguste Picard, 1927), 31. On Islamic textiles in the kingdom of Sicily, see Serjeant, *Islamic Textiles . . .*, ch. 19, 191–2.

16. Robert Bartlett, *The Making of Europe: Conquest, Colonization and Cultural Change 950–1350* (Princeton: Princeton University Press, 1993), 230.

17. Max von Boehn, *Modes and Manners, vol. 1: From the Decline of the Ancient World to the Renaissance,* trans. Joan Joshua (Philadelphia: J. B. Lippincottt, 1932), 186–7.

18. Mane and Piponnier, *Dress in the Middle Ages,* 59.

19. Jennifer Harris, "'Estroit vestu et menu cosu': evidence for the construction of twelfth-century dress," *Medieval Art: Recent Perspectives. A Memorial Tribute to C. R. Dodwell* (Manchester: Manchester University Press, 1998), 89.

20. Zoé Oldenbourg, *The Crusades,* trans. Anne Carter (New York: Pantheon, 1966), 328. Cf. Friedrich Heer, *Kreuzzüge—gestern, heute, morgen?* (Lucerne and Frankfort: Bucher, 1969).

21. François Boucher, *20,000 Years of Fashion: the History of Costume and Personal Adornment,* Rev. English ed. (New York: Abrams, 1987), 170–8.

22. For example *Antioche* 2256–2367; 2641; 3370–3436, Tancred sells booty for food; 3478–3514, inflated food prices and stories of eating boots; 4039–54, where the Tafurs begin eating dead Turks; 5541–2; 6472–4; 6976–96, eating grass and raw donkey; 7110–11, 25 days of hunger; 7275; 9347–60, supply convoy of 500 camels captured; *Jérusalem,* 1295–7, 2432–5, taking supply convoys; 4856–89, 9667–82, looting of food and treasure after battles for Jerusalem.

23. As Orderic Vitalis reports, "Praedia uero hactentus kara uili precio nunc uendebantur," once-valuable possessions were sold cheaply. *Historia ecclesiastica,* ed. Marjorie Chibnall, 6 vols. (Oxford: Clarendon, 1969–1980), 5:16.

24. For citations of original sources see Jonathan Riley-Smith, "The Motives of the Earliest Crusaders and the Settlement of Latin Palestine," *English Historical Review* 98, 389 (1983): 723.

25. Many church relics are obviously of Islamic fabrication. Many sport inscriptions in Arabic. One excellent example of Crusade booty becoming a relic is the "Veil of Sainte Anne," in Apt, Vaucluse, probably booty from the First Crusade. Originally a man's garment, its decorative bands, with a Coptic pattern of roundels probably of later Islamic Iberian manufacture, suggest that it was originally intended as an honorific garment of the first or second grade, marking a warrior or leader's place in the bureaucratic/military hierarchy. Baker, *Islamic Textiles,* 57.

26. Roland Barthes, *Système de la mode,* 10.

27. Duparc-Quioc, *Le Cycle de la croisade,* 47.

28. "François sont covoitos forment en lor corage:/ Se il par covoitise i tornent le visage,/ Dont porés bien poi vaura lo banage./ S'il vienent en estor il i avront damage,/ Car qui est covoitos sovent en a hontage [The Franks are extremely covetous, at heart: if they turn toward us by covetousness, then you can think as little as you want of their right to be here. If they attack us they will suffer losses, for whoever is covetous often comes to shame."] (*Jérusalem,* 7196–7200).

29. Already described and admired earlier, *Jérusalem* 6571–81. Cf. *Antioche* 3032–3116, where the Franks covet the Fabur's charger (guardian of one of the gates of Antioch).

30. There are several places in the cycle (e.g., *Antioche* 4481, 5518–19, 5761) where the narrator makes it clear that items were Islamic in nature, as opposed to the mirror images of Europeans often found in the *chansons de geste* (Bancourt, *Les musulmans dans les chansons de geste,* 2: 580).

31. Cf. the anti-venom belt buckle of the figure of Wealth in the *Roman de la Rose,* ed. Félix Lecoy (Paris: Champion, 1966), 1065–74.

32. For amplification's role in contemporary rhetoric, Edmond Faral, *Les Arts poétiques du XIIe et du XIIIe siècle. Recherches et documents sur la technique littéraire du Moyen Age* (Paris: Champion, 1962), 61–85; Zumthor, *Essai de poétique médiévale,* 51, 85–90. On generic problems of the work, see Robert Francis Cook, *Chanson de Geste: Le Cycle de la croisade est-il épique?* (Amsterdam: 1980).

33. For example the series in *Jérusalem* 7870–80, 7915–18, 7946–52, 7986–9, 8042–7, 8082–6, 8113–17.

34. As seems to have been the case in Europe after the decline of the Roman Empire, when trade and production were extremely limited and the economy had little liquidity. A few elites were able to dress sumptuously, but the mechanisms for constant change and widespread consumption of novelty were not in place.

35. See the numerous excellent essays in Stewart Gordon, ed., *Robes and Honor: The Medieval World of Investiture,* The New Middle Ages (New York: Palgrave, 2001); also Patricia L. Baker, "Islamic Honorific Garments," *Costume* 25 (1991): 25–35; Paula Sanders, *Ritual, Politics and the City in Fatimid Cairo* (Albany: State University of New York Press, 1994); concerning slightly later periods, Thomas T. Allsen, *Commodity and Exchange in the Mongol Empire: A Cultural History of Islamic Textiles* (Cambridge: Cambridge University Press, 1997); Karl Stowasser, "Manners and Customs at the Mamluk Court," *Muqarnas: An Annual on Islamic Art and Architecture* 2 (1984): 13–20; L. A. Mayer, *Mamluk Costume: A Survey* (Geneva: Albert Kundig, 1952), 56–64.

36. Gordon, "Robes, Kings, and Semiotic Ambiguity," in *Robes and Honor,* 383–4.

37. Paula Sanders, "Robes of Honor in Fatimid Egypt," in *Robes and Honor,* 225–39; and Gavin R. G. Hambly, "From Baghdad to Bukhara, From Ghazna to Delhi: The *Khil'a* Ceremony and the Transmission of Kingly Pomp and Circumstance," in *Robes and Honor,* 192–222. Sanders notes that by the tenth century, robes were less often actually cast off, but that those that had actually been worn by a caliph were held to have *baraka,* literally a "blessing," an aura that radiated the caliph's authority as spiritual guide and interpreter of tradition.

38. Yedida Kalfon Stillman and ed. Norman A. Stillman, *Arab Dress: From the Dawn of Islam to Modern Times* (Leiden/Boston: Brill, 2000), esp. 41–60.

Under the Turkish dynasties, 62–85; in the early days of Islam, 7–28, 161–74; *Tiraz* inscriptions in medieval Egypt, which evolved and changed over time, "were admired and desired by almost everyone. . . ." Veronika Gervers, "Rags to Riches: Medieval Islamic Textiles," *Rotunda* 11/4 (Toronto, Royal Ontario Museum, 1978–79): 22–31.

39. Stillman, *Arab Dress,* 107.

40. Lipovetsky argues that a fashion system must be in constant and permanent revolution—if the system ceases, it is not a fashion system *strictu sensu. L'empire de l'éphémère,* 31, 39, *passim.*

41. For example *Antioche,* 396–99, 449–51, 8897–9, and *Jérusalem,* 6302, gift exchange among Saracens; 1054–5, Crusaders receive gifts from the emperor at Constantinople; 4034–6, 4210, 4252, Rainalt Porcet is tempted to convert with gifts and good food; *Jérusalem,* 429–37,472–4, clothes given to Baudoin de Beauvais and Jehan d'Alis by Corbaran in the *Chétifs.*

42. Gifts of various kinds are mentioned: *Chétifs,* 426–33; 556–71; 598–607; 651–7; 665–9; 696; 724–5; 760–80, 785–800, 804–19;1101–4; 1306–7; 1455–6; 2202–7; 2975–9; 3222–3; 3249–55; 3336–40; 3847–55; 3894–96. It is true that the captives are in rags at the beginning of the story, so they could not have fought without some equipment; that need not have been described in detail or mentioned repeatedly, however.

43. Supposedly based on the historical Kerbogha or Karbuqa, Atabeg of Mosul, the Crusaders' worst enemy at Antioch, who disappears from local rulership after his defeat. On this historical figure, Myers, introduction to *Chétifs,* xxii; Steven Runciman, *A History of the Crusades,* 3 vols. (Cambridge: Cambridge University Press, 1951), 246–9.

44. *Bliaut,* a narrow tunic laced on both sides and closed on the breast with a button or brooch. Worn by both sexes from the eleventh century until replaced by the surcot in the thirteenth century. Victor Gay, *Glossaire archéologique du moyen âge et de la renaissance* (Paris, 1887; 1928), 161.

45. Baker, *Islamic Textiles,* 25, 53–63; Stillman, *Arab Dress,* 40–1.

46. In the royal French sumptuary laws of 1279 and 1294, regulations restrict the number of robes given to companions according to income levels. Heller, "Light as Glamour," 948–9; H. Duplès-Agier, "Ordonnance somptuaire inédite de Philippe le Hardi," in *Bibliothèque de l'école des Chartes,* ser. 3, vol. 5 (1854): 176–181; Jourdan, Decrusy, and Isambert, *Recueil général des anciennes lois françaises, depuis l'an 420 jusqu'à la révolution de 1789,* 29 vols. (Paris, 1821–23), 2: 697–700; Eusèbe de Laurière et al, *Ordonnances des roys de France de la troisième race,* 29 vols. (Paris, 1720–23), 1: 541–3.

47. On Frankish stereotypes of the Muslim Pantheon, see Bancourt, *Les Musulmans dans les chansons de geste,* 355–549; Skidmore, *Moral Traits of Christian and Saracen,* 27, 32; Leo Spitzer, "Tervagant," *Romania* 70 (1948–49): 397–407.

48. The Tafurs are a fascinating part of the cycle that deserves more study. They seem to be one of the embarrassing aspects that has kept the works from being considered a "national epic." See Alexander Haggerty Krappe,

"L'Anthropophagie des Thafurs," *Neophilologus* 15 (1930): 274–278; Malouf shows that they are remembered in Arab chronicles, *The Crusades Through Arab Eyes*, 37–55.

49. Marcus Bull, "The Roots of Lay Enthusiasm for the First Crusade," *History* 78 (1993): 353–372, esp. 354–55.

CHAPTER 7

"CHRIST AS A WINDBLOWN SLEEVE": THE AMBIGUITY OF CLOTHING AS SIGN IN GOTTFRIED VON STRAßBURG'S *TRISTAN*

Margarita Yanson

In Gottfried von Straßburg's romance of *Tristan,* written around 1210 and presenting by far the most elaborate version of the legend, the scene of Isolda's ordeal occupies one of the central places. Isolda, the famous medieval adulteress, succeeds in passing her trial by hot iron to the great surprise of King Mark, the bishops, and the courtiers. The author, who never fails to provide his own comments and explanations for the major events in the narrative, attributes the heroine's success to the judgment of Christ—an arbitrator superior to the court of human justice. Gottfried explains that

> dâ wart wol g'offenbaeret
> und al der werlt bewaeret,
> daz der vil tugenthafte Crist
> wintschaffen alse ein ermel ist.
> er vüeget unde suochet an,
> dâ man 'z an in gesuochen kan,
> alse gevuoge und alse wol,
> als er von allem rehte sol

> [Thus it was manifest and confirmed to all the world that Christ in His great virtue is pliant as a windblown sleeve. He falls into place and clings, whichever way you try Him, closely and smoothly, as He is bound to do.][1]

Consequently, Isolda's exoneration results directly from the divine judgment, which, as the author shows, is not only superior to human judgment, but also independent of it. The statement is bold in itself, but Gottfried's comparison of Christ to the most opulent and detachable element of the noble attire produces a striking if not shocking effect on the audience. Why this image? The simile is not merely unorthodox, but ambiguous in its essence. With this image in mind, I would like to consider the symbolic aspect of clothing in Gottfried von Straßburg's *Tristan,* its role as a sign and its deliberate ambiguity as such. Particularly, I am interested in examining the symbolic role of clothing in the context of Gottfried's poetic project as a whole and its relationship to the main themes of the romance.

To emphasize the significance of clothing to the fate of Tristan, I would like to begin my discussion with the dressing trick that contributes to the hero's birth. Tristan's mother Blanscheflur disguises herself as a beggar in order to be able to visit her beloved knight Riwalin. The beggar's rags provide the only possibility for Blanscheflur to see the knight, who at that moment is dying from the battle wounds. The salutary effect of her visit results not only in the rapid recovery of Riwalin, but also in Blanscheflur's pregnancy. For that reason Tristan, the child born to the union of Blanscheflur and Riwalin, owes his appearance in this world to his mother's trick of concealing her identity through "dressing down"—a trick that Tristan inherits and employs so well throughout the romance.

Yet, already in this small episode, we can discern the problematic aspect of dressing, as well as its close association with the main themes of the romance. While in this author's opinion the birth of Tristan presents an unquestionably positive event, in popular view it is tainted by the fact that the child was illegitimately conceived. Morgan, the usurper of Tristan's kingdom, alludes to this fact and contests Tristan's right to inherit his father's land.[2] Blanscheflur's dressing trick results, therefore, in two contradictory effects: the positive event of the hero's birth and the negative implication of his illegitimacy, which carries serious political complications for Tristan. Furthermore, the hero's questionable birth contributes to the ambiguity of his identity. Tristan, indeed, has two fathers and none, as he himself confesses when Raul, his foster father, reveals the secret of the child's true origin at the court of King Mark.[3] By his birthright Tristan is eligible to inherit the kingdom of his father and the kingdom of his uncle, yet he inherits neither one of them.

Similarly, Tristan's own disguises through clothing in many different episodes of the romance serve the double purpose: they emphasize the complexity of his identity while questioning it at the same time. Gottfried presents his hero to the audience now dressed as a courtier, now in a minstrels' costume, now as a merchant, as a royal messenger in Ireland, and as a pilgrim in England. And the reader is constantly challenged to reconcile Tristan's conflicting costume appearances with the singular nature of the hero.

The common bond between Tristan and Isolda reveals itself long before they drink the love portion. Among many other skills that the lovers share is their ability to both veil and unveil their identities through clothing. Even in the episodes that seemingly suggest a rather straightforward association between the their attires and personalities, the function of clothing remains ambiguous. Such is the scene of Tristan and Isolda's procession at the court of Ireland, where for the first time both of them appear fully invested in their royal attires. The scene starts with Tristan's giving gifts of decorative ornament pieces to the three queens of the Irish court. The hero's role as a gift-giver of clothing is unusual; in the medieval courtly culture, and especially in its rendering in the medieval romances, bestowing garments on the high-ranking guests has been a particular duty of the queen.[4] Here we have a reverse situation of a royal male figure offering gifts of clothing to the female royal characters:

> under diu was ouch Tristande
> sîn schrîn und sîniu cleider komen.
> dâ haete er sunder ûz genomen
> drî gürtele den vrouwen drîn,
> daz keiserîn noch künigîn
> nie keinen bezzeren gewan.
> schapel unde vürspan,
> seckel unde vingerlîn,
> der was ebene vol der schrîn
> und was daz allez alsô guot,
> daz niemer keines herzen muot
> des gedenken möhte,
> waz ez bezzer töhte.

> [Meanwhile, Tristan's chest and clothes had arrived. From it he selected a girdle for each of the three ladies, so fine that no queen or empress ever had a better.
> The chest was full to the lid of chaplets, brooches, purses, and rings, all of such quality that in your fondest fancy you could think of nothing finer.][5]

In addition, Tristan chooses items to adorn himself from the same chest and, therefore, establishes a certain unifying principle between the garments of the queens (particularly of Isolda) and his own attire. In this manner, the exchange of gifts manages to accomplish a double task: it alludes to the equality of Tristan and the queen's royal status, and establishes a symbolic connection between the two main characters, who are soon to fall in love with each other on their sea journey to King Mark's court.

The meaning of Tristan's royal representation through his own attire is problematic for several reasons. In the scene of his first public presentation as a royal delegate at the court of Ireland, Tristan is supposed to promote the cause of his uncle, King Mark. His elaborate attire is supposed to signify the wealth of Mark's kingdom and consequently to validate publicly Tristan's promise that the proposed marriage union between Mark and Isolda would increase the Irish queen's social status. At the same time, this is the first instance in the romance when Tristan fully signifies his own royal identity, for prior to this episode he has been known to the Irish as a minnesinger and a knight, accordingly. This double signification produces an obvious and deliberate ambiguity. If Tristan's own status is equal to that of Mark's, should he then be regarded as a message, a messenger, or an independent agent? In addition, we should consider that in spite of Tristan's attempt to pass as a symbolic substitution for the king, he himself is the true performer of the acts required to procure Isolda in marriage. Not only does the hero venture on the dangerous journey to Ireland, but he also kills the dragon, making himself a lawful candidate for the royal marriage through his deed of valor. Tristan's active role stands in contrast to Mark's inactivity, while their equal status makes them both eligible to marry Isolda. The issue of Tristan's symbolic elevation to the royal status through his clothing relates in this way to the broader theme of the king's passivity and the hero's active participation in his fortune.

Certain elements in Gottfried's description of Tristan's garment deserve additional consideration. From the first lines of the passage, even before we learn about the details of the hero's dress, the author underlines the harmonious, ideal correspondence of Tristan's dress to his knightly vocation:

> des dinc was ouch ze prîse
> und ze wunder ûf geleit
> an iegelîcher saelekeit,
> diu den ritter schepfen sol.
> ez stuont allez an im wol,
> daz ze ritters lobe stât.
> sîn geschepfede und sîn wât
> die gehullen wunneclîche in ein.
> si bildeten under in zwein
> einen ritterlîchen man

[He was marvelously blessed with every grace that goes to make a knight: everything that makes for knightly distinction was excellent in him. His figure and attire went in delightful harmony to make a picture of chivalrous manhood.][6]

Here the statement of harmony in Tristan's appearance goes beyond the normal correspondence between the attire and the social status, since "ein ritter" is not merely a noble man, but a bearer of an ethical and courtly code of values. Gottfried presents Tristan to both the internal audience of his romance and the external audience of the book as an ideal knight, whose outward excellence matches his inward merit. The authorial statement about the integrity of Tristan's appearance and character does not, however, resolve the problem of the ambiguous message, implicit in the hero's attire. On the contrary, the further complication arises from the contradiction between this statement of harmony and the ambiguity of Tristan's attire as a medium of representation.

Small decorative elements of Tristan's dress echo the ornaments of Isolda's costume. Thus in lines 11114–15, the author mentions "ein netze daz was ûz daz dach/ von cleinen berlîn getragen [a net of tiny pearls]" that Tristan wears and that previously appear in the description of Isolda's mantle. In addition, both Tristan and Isolda wear precious stones, which medieval lapidary tradition associates directly with the corresponding human virtues. Even if a net of small pearls is a standard feature of the noble attire, including this decorative motif in the description of both Tristan and Isolda is emphatic: it accentuates the connection between them even further. Gottfried's acknowledgment of his two protagonists' inward perfection resonates directly with his concept of the "edele herzen [the noble hearts]" that he introduces early in the prologue, and which Walter Haug considers as one of the key notions to the understanding of the author's project at large. According to the critic, the author relies on the nobility of his audience's hearts for the correct interpretation of his *Tristan* and for the proper appreciation of his poetry.[7]

Gottfried's description of Isolda's dress in the same episode occupies 121 lines[8] and presents one of the lengthiest and highly detailed catalogues on fashion. In his depiction of the heroine's garment, the author particularly stresses the idea of proportion and measure, personified in this passage, as one of the main features of Isolda's dress:

si truoc von brûnem samît an
roc unde mantel, in dem snite
von Franze, und was der roc dâ mite
dâ engegene, dâ die sîten
sinkent ûf ir lîten,
gefranzet unde g'enget,
nâhe an ir lîp getwanget
mit einem borten, der lac wol,
dâ der borte ligen sol.

der roc der was ir heinlîch,
er tete sich nâhen zuo der lîch.
. . . .
ern wasze kurz noch ze lanc.
er swebete, dâ er nider sanc,
weder zer erden noch enbor.
dâ stuont ein höfscher zobel vor
der mâze, als in diu Mâze sneit,
weder ze smal noch ze breit,
gesprenget, swaz unde grâ.

[She wore a robe and a mantle of purple samite cut in a French fashion and
accordingly, where the sides slope down to their curves, the robe was fringed
and gathered into her body with a girdle of woven silk, which hung where
girdles hang. Her robe fitted her intimately, it clung close to her body, it neither bulged nor sagged but sat smoothly everywhere all the way down. . . .
For length it was just right, neither dragging nor lifting at the hem. At the
front it was trimmed with fine sable cut to perfect measure as if Lady Mod-
eration herself had cut it, neither too broad nor too narrow][9]

The fact that her dress is made to measure can be interpreted as the queen's
fashion statement only in part. The author indeed takes a special pride in the
French cut of Isolda's dress. Yet the notion of measure indicates not only the
heroine's commitment to courtliness, but also points to the integral proportion within her character. Moreover, the measured opulence of Isolda's attire,
so lovingly rendered and so carefully impressed by the author on his audience's memory, will later provide a contrasting background for Isolda's appearance in a short woolen skirt over the hair shirt during the trial episode.
    But perhaps the most striking feature of Isolda's attire remains its multi-
farious quality. The colors of her dress are also fashionable: brûn (e.g., dark
blue) samit contrasts with the whiteness of ermine and the black and gray
shades of sable; emerald and jasper, sapphire and chalcedony stand against
the golden background of her headdress. Gottfried makes the heroine's attire so luxurious and so visually striking that it creates the effect of a human
inability to cope with this overwhelming display. The description of the lining of Isolda's dress is mystifying: it appears and disappears as if the sight of
it were almost impossible to grasp—now you see it, now you don't:

man sach ez inne und ûzen
und innerthalben lûzen
daz bilde, daz diu Minne
an lîbe und an dem sinne
so schône haete gedraet.

[One saw it inside and out, and—hidden away within—the image that Love had shaped so rarely in body and spirit!][10]

The effect of almost complete blurring is achieved in the description of the gold crown on Isolde's golden hair: "dâ lûhte golt unde golt, der cirkel unde Isolt in widerstrît ein ander an [Gold and gold, the circlet and Isolda, vied to outshine each other.]"[11]

This short passage contains, in my opinion, two extremely important notions. First of all, it provides a context of an absolute unity between the heroine and the element of her attire, which can be taken to stand for her entire costume, based on Gottfried's emphasis on the Right Measure as a linking principle between the heroine and her gown. Later in the romance, in the *Rotte und Harpfe* episode, the author explores this previously established principle of unity. Identification of the heroine with her attire enables Tristan to claim Isolda for his prize from the Irish knight and allows the hero to call his beloved "die aller besten wât [the best dress]" that could be found in the tent.

The second and the more important idea expressed in the passage deals with the visual imperceptibility of Isolda's dress. Together with the notion of unity it reveals Gottfried's conception of Isolda as a figure that is beyond human understanding and, consequently, beyond human judgment. Isolda's golden hair is as indistinguishable from her crown as a pearl on a white forehead ("perla in bianca fronte"), which Dante employs as a metaphor for Paradise in his *Commedia*. The barely distinguishable image of the white pearl allows Dante to demonstrate how impenetrable Paradise is to human comprehension.[12] According to the Aristotelian scheme of knowledge acquisition, dominant throughout the Middle Ages, sense perception plays a constitutive role in the process of cognition. Moreover, Aristotle gives vision priority over other senses. The ability to see correctly determines, therefore, the ability to know and, consequently, to make an accurate judgment. Isolda's dress with its various intermingling colors, the ever-escaping sight of her lining and the crown that blends with the heroine's hair produces such an overwhelming, over-saturated image, that human eyes fail to grasp the entire picture at once. In this way the author suggests that in order to understand and to judge Isolda, the audience can rely neither on the physical reality of the world, nor on the human capacity to perceive. Instead, the readers are invited to grasp the inner sense of the represented object through the internal vision of their hearts. This concept, so resonant with Dante's treatment of vision in *Paradiso,* refers us back to Gottfried's prologue and its notion of the noble heart as an instrumental characteristic for the true judgment of *Tristan* again. Seen in this way, the ambiguity of Isolda's dress as a sign results from the essential contradiction between inability to perceive it and the necessity to perceive in order to be able to understand the true character of the heroine and the meaning of the romance as a whole.

While the scene of *Wahrzeichen* demonstrates Tristan and Isolda's ability to use clothing as a means of self-representation, other episodes in romance exhibit the couple's ability to construct false identities and handy lies through dressing tricks. When Tristan first comes to Ireland to be cured from his mortal wound, he presents himself as a minnesinger. The hero signifies the change of his identity through three particular elements: he changes his name, he changes his dress, and, finally, he leaves all of his possessions with the exception of the harp behind:

> Tristan ime dô geben bat
> daz aller ermeste gewant,
> daz man in der barken vant.
> und als man ime daz ane getete,
> er hiez sich legen an der stete
> ûz der barken in daz schiffelîn.
> sîne harpfen hiez er ouch dar în
> und in der mâze spîse geben,
> daz er ir möhte geleben
> drî tage oder viere.

[. . . he[Tristan] asked them for the very worst clothes in the barque. When they had dressed him in them, he ordered them to remove him from the barque without delay and place him in the skiff. He also told them to place his harp inside it and food enough t sustain him for three or four days.][13]

From that moment on the harp becomes an essential element in Tristan's ploy to pass for a minstrel. It acquires, therefore, a symbolic status and signifies Tristan's transformation. However, Tristan's newly assumed identity does not completely negate or even contradict his true self. His assumed name "Tantris" is merely "Tristan" reversed; Tristan's skill at harping is part of his courtly upbringing that has already won him renown at the court of Mark. Similarly, his poor attire is associated directly both with his mother's disguise as a beggar during her visit with Riwalin, and with Tristan's elusive role of a pilgrim in the world. When Tantris tells the story of his plight to the Irish audience, he claims not only to have spent 40 nights and 40 days at sea (7598)—a double allusion to Christ's 40-day fasting in the desert and the story of Christ and his disciples at sea during the storm; he is also called "ein armer marteraere" (7648), which can be both translated as "a poor sufferer" and "a poor martyr." Nevertheless, it would be wrong to say that Gottfried's appeal to the Gospel stories is meant to establish a direct parallel between his romance and the New Testament. Rather, this indirect allusion broadens the framework of the narrative. Tristan, who is seeking a cure for his mortal wound at that point in the romance, is also a sufferer who is seeking salvation in the broader Christian context. The more concrete meaning of the events within the narrative evokes in this fashion an association with the more abstract and universal condition of humanity at large. Tristan's poor cloak is meant both to deceive the Irish court and to draw the reading audience to the broader context of the romance.

Even more deceptive is Tristan's appearance in a pilgrim's costume at the shore of England, before Isolda's trial. He disguises himself, following Isolda's request. The pilgrim's cloak becomes part of Isolda's trick to simulate the situation in which she could openly lie in Tristan's arms:

Nu diz geschach. Tristan kam dar
in pilgerînes waete.
sîn antlütze er haete
misseverwet unde geswellet,
lîp unde wât verstellet.

[This was duly done. Tristan repaired there in pilgrim's garb. He has stained and blistered his face and disfigured his body and clothes.][14]

During her trial by hot iron, this trick enables the heroine to tell the truth, which is also a lie, and a lie, which is also true. Here again, Isolda's ability to manipulate the situation to her advantage does not rely entirely on the outward representation. Tristan's pilgrim's garb is only partially misleading: it helps to fool Mark and the court of England, but it does not contradict the essential quality of Tristan's character. Isolda's deceit is partly justified, because it parallels Mark and the court's own self-deception: they are looking for the outward sign of the queen's guilt while being blind (on the account of their defective perception) to the message of the inward love between Tristan and Isolda. They do not possess the noble hearts to see beyond the obvious.

In this scene, Tristan's disrobing of his noble attire corresponds to Isolda's consequent divestment of her opulent dress. In this manner the author preserves the previously established parallel between their costumes. If during the *Wahrzeichen* episode the analogy between Tristan and Isolda's attires serves the double purpose of demonstrating their equality in status outwardly and establishing the inward connection between the heroes, then in the case of the ordeal episode it provides a further insight into the unity between the two lovers. Tristan and Isolda are joined not only by their conspiracy to dupe the court, but also by their mutual plight. After all, Isolda's failure to prove her innocence would carry the same implication for her lover Tristan.

Moreover, Tristan's disguise as a pilgrim alludes once again to the broader context of the romance. Only this time the allusion is much stronger, since the whole disguise episode occurs in anticipation of the judicial trial. As the sight of Isolda's luxurious dress during the procession in Ireland escapes human judgment and stands in the way of assessing her true identity, likewise Tristan's pilgrim attire escapes the understanding of the characters within the narrative. Instead, the message of his dress is directed to the external audience and informs it about the hero's standing with respect to the transcendental reality. It is from that perspective that Isolda might very well be guiltless, and Tristan could easily be a pilgrim. The ambiguity of Tristan's pilgrim cloak as a sign results, therefore, from its ability to trick and to reveal, to lie and to tell the truth simultaneously. Its function as a sign is deliberately equivocal, due not only to the limitations of the human capacity to judge, but also because the author purposely complicates the implicit message by addressing it to two distinctly different audiences: one within the narrative frame of his romance, and the other outside the fictional reality of his work.

The exceptional attention that Gottfried pays to clothing leads some critics to consider the author's discussions of attire as a method in switching narrative modes. In her study on the role of clothing in *Tristan,* Gabriele Raudszus notices that whenever Gottfried treats the character's attire as an allegory, he switches from glorification of their noble status to the issues of ethics.[15] As an example of Gottfried's use of allegory, the critic cites the Bragane episode, in which the bloody cloth presented to Isolda by the hunters stands for the corresponding bloody act of murder. The obvious problem with this example, in my opinion, is exactly the failure of allegory to represent the signified correctly; for we know Bragane to be alive at this point of the story, and the smeared garment to be a mere deception—a signification of something that has never happened. The allegory in this episode is, therefore, deliberately misleading: the bloody cloth stands for nothing, for deceit.

The *Rotte und Harpfe* episode provides a more characteristic use of allegory by the author. In this episode Tristan and his rival, the Irish knight, are represented allegorically though their musical instruments, and Isolda is also referred to as the best attire that Tristan is able to find in the Irish knight's tent. After Tristan tricks the knight and manages to ride away with Isolda, the Irish knight recognizes his defeat in the following terms:

"vriunt . . . ir gebt rîlîche wât.
ich hân daz beste gewant,
daz ich in dem gezelte vant!"

["You have my word on it, my friend," replied the Irishman . . ."I will give
you the finest clothes that we have in this pavilion."][16]

Tristan tricks the trickster using the same ruse employed by the Irish
knight himself employs to win Isolda from King Mark. The entire development of the plot within this scene is constructed around the gradually
conveyed meaning of allegory. By the time king Mark realizes that a gift
he promises to the foreign knight in return for his musical performance
can stand not only for an object, but also for his wife, it is too late for him
to reverse his rash promise. Likewise, the knight is incapable of conceiving
that a prize of a dress can be interpreted as the actual bearer of the attire.
Both the king and the knight exhibit a limited perception of the situation.

Mark's superficiality is traditional to his character, and it comes as no
surprise that he is incapable of graduating to the higher level of insight.
Even after the Irish knight reveals the meaning of the allegory to the king
and claims Isolda for his prize, Mark remains clueless as to the seriousness
of the political implications the loss of the queen imposes on the kingdom.
It takes Tristan to disclose the full meaning of Mark's mistake:

»hêrre« sprach er »wizze crist,
sô lieb als iu diu künegîn ist,
sô ist ez ein michel unsin,
daz ir si gebet sô lîhte hin
durch harpfen oder durch rotten.
ez mac diu werlt wol spotten.
wer gesach ie mêre künigîn
durch rottenspil gemeine sîn?
her nâch sô bewâret daz
und hüetet mîner vrouwen baz!«

["Sire," he said, "dear as the Queen is to you, Heaven knows, it is a great
folly on your part to give her away so lightly for the sake of the harp or the
rote. People may well scoff. Whoever saw a Queen made common property
for a performance on the rote? Don't let it happen again, and guard my lady
better in future!"][17]

It is particularly interesting that while admonishing Mark, Tristan also tries
to give him a lesson in allegorical interpretation: he beseeches the king
never to lose his wife either through harping or through rote-playing. Yet,

in this lesson, there is another hidden trick that Mark with his usual ob-
tuseness does not get. It is clear even to the king that the rote stands for
the Irish knight, but this is not the only musical instrument mentioned by
Tristan. The harp comes to be closely associated with the hero himself as
the scene of his first appearance at the court of Ireland reveals. Therefore,
by referring to the harp, Tristan, in fact, informs Mark about his own connection to Isolda and the dangers that this involvement between the king's
nephew and his wife presents to the king. Tristan goes even further and accentuates his message at the end of the speech by calling Isolda his lady,
which in this particular context is different from the usual polite address
to the queen by her subject.

The situation with the Irish knight is somewhat different, for he is obviously more capable of insight into the allegorical meaning than the king.
At the same time, Tristan's victory over him suggests that the knight's insight is also limited, for it is not based on the same intrinsic inward connection as the one two lovers enjoy. Thus the only reason that the knight
becomes vulnerable to deceit is, again, the deficiency of his perception.
The knight invites Tristan to play the harp in order to assuage the misery
of Isolda, who at that point in the story is crying bitterly over her plight.
The tears of Isolda are, therefore, as much a part of Tristan's trick as his disguise as a pilgrim is a part of Isolda's ruse in the ordeal scene. It goes without saying that Isolda's crying is not staged, and the tears that she drops are
completely honest. But at the same time her crying is instrumental in Tristan's ability to manipulate the situation. The Irish knight, on the other
hand, has no knowledge of the connection between Tristan and Isolda. For
what we know, he is likely to believe that Isolda's crying is on account of
her parting from Mark. The knight is therefore no more capable of recognizing true love than the king. Since he does not possess the noble heart,
he is doomed to have a limited understanding and lose Isolda to Tristan,
no matter how clever he is at tricks are otherwise.

However central the allegory is to the *Rotte und Harpfe* episode, it is
in the *Gottesurteil* scene that the author reveals it's full significance to his
romance. Here Isolda's changing of her attire acquires a truly symbolic
meaning; here Gottfried introduces his most striking metaphor of
Christ as a windblown sleeve. It is worth emphasizing that Isolda's voluntary divestment of her elaborate attire is as public as her appearance
in a luxurious garment at the court of Ireland. Furthermore, comparison and contrast between the episodes can take place, since they are the-
matically connected by the judicial purpose of the two gatherings. The
heroine's stripping the rich decorations of her dress and distributing
them among the poor mirrors the splendid spectacle of Isolda's appearance at the Irish court:

daz îsen daz was în geleit.
diu guote küniginne Îsolt
diu haete ir silber unde ir golt,
ir zierde und swaz si haete
an pferden unde an waete
gegeben durch gotes hulde,
daz got ir wâren schulde
an ir niht gedaehte
und sî z'ir êren braehte.

[The good Queen Isolde had given away her silver, her gold, her jewelry, and all the clothes and palfreys she had, to win God's favor, so that He might overlook her very real trespasses and restore her to her honor.][18]

The symbolic meaning of Isolda's act uncovers itself explicitly in this quote. Isolda rids herself of all the signs of her royal status for the sake of establishing a closer connection with God and procuring His benevolence. Two examples from the sources vaguely contemporary to *Tristan* (one literary and one historical) can help us appreciate the significance of Isolda's divestment before her procession to the church. In *Nibelungenlied* Krimhild and Brunhild, the two queens, use the opportunity of going to church in order to compete with each other in the splendor of their garments.[19] At the same time, the vita of St. Elisabeth of Thuringia, another historical royal figure, pictures her as having a custom of detaching her sleeves and taking off her crown in church.[20] For she believes it proper to signify her humility in front of the King of Heaven, whose head is adorned with the crown of thorns. Under normal circumstances Isolda would have been likely to follow the pattern of the romance characters, but during the time of trial she reverts to the practice of the royal saint figure.

Isolda's next "step of humility" is marked by her taking off the sleeves and putting a hair shirt and a short woolen tunic on. Barefoot, with bare arms, Isolda proceeds to the church, evoking compassion in the hearts of many bystanders:

si truoc ze nâhest an ir lîch
ein herte hemede haerîn,
dar obe ein wullîn rockelîn
kurz und daz mê dan einer hant
ob ir enkelînen want.
ir ermel wâren ûf gezogen
vaste unz an den ellenbogen.
arme und vüeze wâren bar.
manec herze und ouge nam ir war
swâre unde erbermeclîche.

[She wore a rough hairshirt next her skin and above it a short woolen robe which failed to reach to her slender ankles by more than a hand's breadth. Her sleeves were folded back right to the elbow; her arms and feet were bare. Many eyes observed her, many hearts felt sorrow and pity for her.][21]

Thus Gottfried completes his outward representation of the heroine's inward transformation. The highly ceremonial, public nature of Isolda's divestment gives her act a symbolic status: by renouncing the attributes of her noble pride, Isolda exempts herself from the court of human judgment and submits herself to the judgment of Christ exclusively.

Detachment of the sleeves deserves a further consideration, because it occurs in the direct anticipation of the author's digression that culminates the Ordeal episode.

In it the poet compares Christ to a windblown sleeve—a metaphor both remarkable and, to my knowledge, unique:

dâ wart wol g'offenbaeret
und al der welt bewaeret,
daz der vil tugenthafte Crist
wintschaffen alse ein ermel ist.
er vüeget unde suochet an,
dâ man'z an in gesuochen kan,
alse gevuoge und als wol,
als er von allem rehte sol.
erst allen herzen bereit,
ze durnehete und ze trügenheit.
ist ez ernest, ist ez spil,
er ist ie, swie sô man wil.

[Thus it was manifest and confirmed to all the world that Christ in His great virtue is pliant as a windblown sleeve. He falls into place and clings, whichever way you try Him, closely and smoothly, as He is bound to do. He is at beck of every heart for honest deeds or fraud. Be it deadly earnest or a game, He is just as you would have Him.][22]

The close association of Isolda with her attire throughout the entire romance on the one hand and the quid pro quo substitution of her elaborate sleeves with Christ on the other, confirms in the eyes of the audience her superior connection to God. Gottfried's lyrical digression about Christ at the end of the episode is entirely for the benefit of the audience, since the readers, unlike the characters in the romance, are fully aware of Tristan and Isolda's adulterous affair. For them the mere sight of the heroine's successful passage through the ordeal would not be enough to justify her innocence. The judgment of Christ, however, provides the necessary proof.

The external state of affairs in the world that is "verkêret [over-
turned]," where "daz honegende gellet, daz süezende siuret, daz touwende
viuret, daz senftende smerzet" [honey changes to gall, sweetness to sour,
fire to moisture, balm to pain][23] does not interfere with the Tristan and
Isolda's inner connection to God through the love of their noble hearts.
The judgment of their love belongs to God alone; the appreciation of
Gottfried's romance is open to readers with noble hearts. But in an up-
side-down world there can be no measure that would allow to distinguish
lie from truth. Therefore, the signs in this world are unreadable, the allegories are ambiguous, and the symbols are obscure. By creating an image
of Christ as a windblown sleeve that attaches itself to any individual need
and purpose, regardless of the human notions of justice, the author challenges the idea of a fundamental connection between the human and the
divine realms, which is the basis of providential history. According to Gottfried, in a dubious world two different names Tristan and Tantris can signify one person, while two different heroines can share the same name of
Isolda. The same man can assume the role of a merchant, a minnesinger,
a pilgrim, a knight, and a king; the same woman can appear to be an adulteress and a saint.

Notes

1. *Tristan* (Stuttgart: Philipp Reclam Jun., 1996), 2, 15733–40. Translation by
   A. T. Hatto. *Tristan with the 'Tristan' of Thomas* (Penguin Books: New York,
   1967), 248.
2. *Tristan*, 1: 5385–444.
3. *Tristan*, 1: 4362–77.
4. A vivid example of the queen's prerogative to bestow gifts on the courtiers
   can be found in *Nibelungenlied*. Krimhild outfits the whole Burgundian
   court with glamorous clothes from her chest. See *Das Nibelungenlied. Ku-
   drun*, ed. Werner Hoffmann (Darmstad: Wissenschaftliche Buchgesellschaft,
   1972), 347–69.
5. *Tristan*, 2: 10816–28; Hatto, 184.
6. *Tristan*, 2: 11092–101; Hatto, 187.
7. Walter Haug. *Vernacular Literary Theory in the Middle Ages. The German Tradition, 800–1300, in its European Context* (Cambridge University Press:
   Cambridge, 1997), 209–11.
8. *Tristan*, 2:10900–20.
9. *Tristan*, 2: 10900–10, 10924–7; Hatto, 185.
10. *Tristan*, 2: 10949–53; Hatto, 185.
11. *Tristan*, 2: 10977–9; Hatto, 186.
12. Dante Alighieri, *Divine Comedy. Paradiso,* Canto 3: 14 (New York, Toronto,
    london, Sydney, Auckland: Bantam Book).
13. *Tristan*, 1: 7420–9; Hatto, 140.

14. *Tristan,* 2: 15560–4; Hatto, 246.

15. In her book on sign language of clothing, Gabriele Raudszuz writes, "Wo die Kleidung nur noch als Symbol fungiert, ist der Übergang von der Deskripzion zur Allegorese vollzogen, läßt sich daran die Tendenz des Dichters zur Ethisierung seines Stoffes und zur Abwendung von der Verherrlichung der Statussymbole ablesen" in *Die Zeichensprache der Kleidung* (Hildesheim, Zürich, New York: Georg Olms Verlag, 1985), 154.

16. *Tristan,* 2: 13420–2; Hatto, 216.

17. *Tristan,* 2: 13440–5; Hatto, 218.

18. *Tristan,* 2: 15641–50; Hatto, 247.

19. See *Nibelungenlied,* Âventiure 14, especially 831–8.

20. See Dietrich von Apolda. *Die Vita der heiligen Elisabeth,* ed. Monika Rener (Marburg : N. G. Elwert Verlag, 1993).

21. *Tristan,* 2: 15656–65; Hatto, 247.

22. *Tristan,* 2: 15733–44; Hatto, 248.

23. *Tristan,* 2: 11884–9; Hatto, 198.

CHAPTER 8

ADDRESSING THE LAW:
COSTUME AS SIGNIFIER IN
MEDIEVAL LEGAL MINIATURES

Susan L'Engle

Much of what we know of the history of costume has been compiled from depictions of clothing and accessories in works of art from all periods and in all media, representing religious and secular themes. A little-explored avenue has been the function of garments and textiles in expressing the law, as found in illustrations to thirteenth- and fourteenth-century manuscripts of Roman and canon law. Here illuminators manipulated elements of fashion, fit, accessories, and hairstyle, to characterize passages that discuss, for example, the division of authority, marriage contracts, crime and punishment, or the proper behavior of laymen and clerics. The physical appearance of human beings and their surroundings in these pictorial compositions informs us about contemporaneous aesthetic conventions, and this visual construction of cultural identity gives us an idea of how the viewer was expected to react to the image. Along with their often ingenious interpretation of textual themes, legal miniatures cast some light as well on the ways in which artists visually presented complex juridical concepts to medieval viewers. This chapter will explore three basic levels at which textiles and dress are used in a legal context: first, as background scenery; second, to identify protagonists, and last, to connote or explicate a point of law.

Textiles as Scenery

Legal rituals and ceremonies, court trials and judgments often take place in elaborately staged scenarios, where the use of textiles may create an

appropriate ambience, identify the protagonists and establish their hierarchy, and provide ceremonial props. Essential to the legal scene is an enthroned figure representing authority—secular, ecclesiastic, or divine. In
manuscripts of canon law this figure is usually a pope or bishop in ecclesiastical dress; the images of Christ or God the Father can be substituted to convey supreme or divine authority. The Emperor Justinian,
considered the father of civil law, presides in Roman law manuscripts although he appears in various guises: clothed in armor as a Roman warrior; wearing luxurious garments and the golden crown of an emperor;
or clad in the fur-collared and -trimmed robes and black *biretta* of a civil
lawyer. In addition, like the cloth of honor stretched behind holy figures
in religious compositions, a richly patterned curtain or backdrop imitating costly materials may serve as background for the authority figures,
emphasizing the importance of their rank and office and the activities
that take place under their supervision. In Figure 8.1, a large two-column
miniature opening a fourteenth-century *Decretales,*[1] pope Gregory IX in
elaborate robes is seated on a throne, feet resting on a puffy banana-
shaped cushion upholstered in patterned cloth. The length of gold-
trimmed brocade suspended behind distinguishes him as official
purveyor of the law. A fourteenth-century illustration to the *Liber Sextus,*
reflecting the teaching motifs so popular at this time in Bologna, depicts
the diminutive Christ Child addressing the Doctors in the Temple,[2] spot-
lighted by a cascade of golden cloth whose delicate leafy designs are
echoed in the shimmering background above, where angels hover. The
patterns and motifs copy those of precious fabrics imported from abroad,
destined for the garments of the wealthy or for decorating the religious
institutions of their patronage. Cost and rarity of material are thus
equated with power and professional status. Furniture associated with
these authority figures, such as thrones and altars, is also draped with precious cloths. A favorite artistic device is to clothe clerics in the same type
of material that covers an altar or a pulpit, partly a decorative ploy, but
by association imparting to all a special degree of sanctity. One of the illuminators of an early fourteenth-century *Decretum Gratiani* in Siena[3] is
distinguished by the wide range of patterns he employs to embellish his
painted clothing and draperies.

Marriage and its associated nuptial rites provided another excuse for
concocting an elaborate visual scenario, although the resulting compositions do not always correspond with historical reality. Illustrating sections
of text devoted to church and civil regulations on marriage, most thir-
teenth- and fourteenth-century miniatures feature a ceremony presided
over by a secular or ecclesiastical authority, depicting the joining of the
bride and groom's hands, and/or the groom's presentation of a ring to the

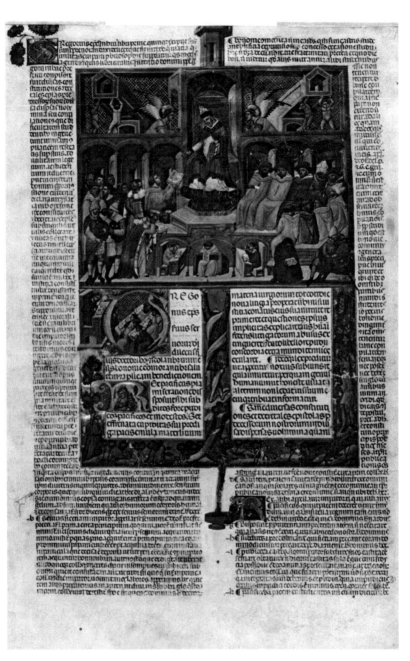

Figure 8.1 Gregory IX, *Decretals:* Prologue, 1330s. New York, The Pierpont Morgan Library. MS M.716, fol.1. With permission of The Pierpont Morgan Library.

Figure 8.2   Gregory IX, *Decretals:* Book 4, ca. 1300. Cambridge, Fitzwilliam Museum, University of Cambridge. MS. McClean 136, fol.188. With permission of Fitzwilliam Museum, University of Cambridge.

bride, as a pledge of love and fidelity. Although in many cases this ritual takes place without reference to physical location, under Italian custom it would generally take place in the house of the bride, while in northern countries it was held outside, in public, and in front of a church.[4] In the fourteenth century, responding to increasingly successful Church efforts to incorporate the marriage ceremony, illuminators began to stage it *within* the church, conducted by a cleric near an altar. A crowded composition in the miniature for Book 4 of a *Decretales* manuscript in Cambridge[5] (Figure.8.2) executed by a Northern European artist and dating to around 1300, depicts the richly attired wedding couple kneeling before an equally splendid tonsured priest in ceremonial robes. Two acolytes isolate this scene from the wedding party and other attending clerics by stretching around it a length of gold brocade, probably representing the *velamen celeste*—the Veil of Heaven. Of varied historical traditions, the suspension of this veil, or of a canopy, at the moment the marital union is blessed—symbolizing celestial approval—was established in Christian marriage liturgy in the ninth century, and also later included in Gratian's twelfth-century *Decretum*.[6] Around 1340 a Bolognese artist concocted a very elaborate narrative version for another *Decretales* in the Vatican Library,[7] staging it in three consecutive episodes and locating all three within a chapel. The ceremony of betrothal is enacted at the extreme left, where the groom places a ring on the bride's finger under the eyes of a secular judge. The ritual blessing takes place before an altar decked with a richly gilded altarpiece, and here a tonsured priest holds an open book and pronounces the appropriate words over the same young couple. Instead of a Veil of Heaven, the wedding couple is wrapped in a poncho-like garment, probably symbolizing, with its single neck hole, the union of two into one. This scene is attended by the bride and groom's parties, and, in addition, closely supervised by a secular judge. At far right the fruit of the marriage rites is displayed, as a distinctly *enceinte* bride exits the scene, one hand laid delicately over her straining belly. The entire narrative is both set off and unified by the gold brocade curtain that runs the width of the chapel.

Clothing the Protagonists

We now turn to the garments themselves, and their function in the construction of a visual identity. Obviously human figures have to be clothed, but costume in legal manuscripts can dictate to the viewer the social, moral, and legal condition of each protagonist in a juridical scene. At a first level, costume is used to establish a social and professional hierarchy among the figures in a composition. The most elaborate clothing is reserved for the authority figure who regulates the legal scene, whether emperor, bishop, or secular judge.[8] When this figure appears in the company of others, he usually occupies the most prominent space, as does Pope Gregory in Figure 8.1, located at center atop a multileveled platform, seated under a baldachin, and framed by an architectural niche. He is resplendent in tiara and papal robes, trimmed with embroidered borders. The costumes of the ecclesiastic dignitaries seated below Gregory at right are less elaborate, and their lower hierarchical ranks are distinguished by specific headgear: the distinctive red hats of cardinals, and the twin-peaked episcopal mitres. While churchmen were generally depicted in their ceremonial robes of office, the garments of secular legal personnel had more variety. Civil officials in Figure 8.1 stand at lower left: lawyers and judges are identified by their fur-lined robes, and wear on their heads fur-trimmed hats or the flat black *biretta;* the military forces are outfitted in chain mail and armor. Secular court scenes were often presided over by the emblematic figure of the Emperor Justinian, embodying ruler, lawyer/judge, or soldier, and sometimes all at once, as exemplified in a fourteenth-century Bolognese miniature opening the Roman law *Institutiones.*[9] Here the crowned emperor holds an upright sword that symbolizes the military enforcement of justice, and wears a scarlet miniver-lined *tabard,* or judge's robe, topped by a scarlet fur-lined shoulder piece. He is surrounded by other court officials at lower levels attired in an assortment of fur-lined or -trimmed robes, complemented by fur-trimmed hats or hoods of various shapes, some in turban style. Variant opening compositions to the *Institutiones* interpret a phrase from the Prologue to Book 1: *Imperial Majesty should not only be graced with arms but armed with laws,*[10] representing a military defense of the law. In these examples Justinian's garments may copy the uniform of a Roman warrior, whose components include a leather cuirass worn over a woolen tunic, a skirt of leather strips stretching from waist to mid-thigh, a pleated leather doublet with fringed edges, and molded shoulder straps. The emperor may carry a sword and shield, and wear a piece of armor protecting the right leg. In another typical trial scene[11] we see the emperor actually judging a court case: in this composition the crowned Justinian wears an unlined *tabard* over his *tunica* and the accompanying lawyers are identified by the full white coif tied under the chin and covering the ears. The great variety of headgear for judges included a rounded or flat *biretta* or *pileus* worn over the coif, sometimes with a twig-shaped topping, although legal figures were often depicted bareheaded.

If we take many of these miniatures at face value, it would seem that medieval law courts were consistently headed by wealthy and upper-class personnel, since luxury trimmings such as fur linings automatically evoke status and power. This is not necessarily the case. We must be aware that these compositions are products of social attitudes and intentions, "constructs," as

Figure 8.3   Justinian, *Codex:* Book 6, fourteenth century. Paris, Bibliothèque nationale de France. MS Latin 14339, fol. 203. With permission of Bibliothèque nationale de France.

Jonathan Alexander has put it, "[not to] be seen, for all their 'realism,' as neutral,"[12] and charged with the ideology their creators wish to express. Whether all lawyers and judges were moneyed or commanded high salaries is moot—what we *can* conclude from their rich costume and privileged roles in the miniatures, is that the process of law and the individuals who enforced it were held in high regard in contemporary society, and it was thought necessary to advertise this fact. Likewise, the less-esteemed members of lower social professions are identified by more modest clothing and less refined physical attributes, as will be demonstrated further on.

In addition to the central authority figure whose function is to discuss points of law, rule on a case, or sentence a lawbreaker, and the lawyers and jurists who present and argue cases, a ubiquitous presence in legal miniatures is the scribe or notary who records proceedings, representing the official and the legitimate. His distinct costume and posture make him instantly recognizable: up to mid-fourteenth century he wears a shapeless, flowing robe, and often a long pointed hood, the end of which is wound around his head to form a floppy turban and knotted at one side. Generally placed in the foreground, sometimes with one leg crossed over the other, at times with inkpot in hand and carrying a pouch filled with extra quills, he busily writes on a scroll or codex. His presence as record-keeper symbolically validates the text he precedes and endorses the activities of the lawmakers and enforcers. In Figure 8.1 a trio of scribes pursue their office from little niches directly beneath the central platform upon which Gregory holds court; in Figure 8.3 a single figure sits at the judge's feet, documenting arguments and discussions such as the accusation of theft taking place immediately to the left. In later fourteenth-century miniatures this indispensable professional is pictured at work before a special writing table like those seen in Figure 8.1, and the increased status he has achieved in society is demonstrated by the fur-lined hat he wears in replacement for the previous turban.[13]

Costume and Society

The human figures present in legal illustrations come from assorted professions and social levels. Rural inhabitants—hunters and agricultural workers, land surveyors, carters, and shepherds—appear in miniatures that deal with purchase, sale, or inheritance of land and animals, and the usufruct of rural products and services. Staged in an outdoor locale, the compositions are populated by individuals with bare legs and feet, wearing short robes and sometimes large and amusing hats. In a miniature illustrating a civil law case dealing with contracts of purchase and the terms settled between vendor and purchaser,[14] (Figure 8.4) the rural context is

Figure 8.4   Justinian, *Digestum vetus:* Book 18, 1340–50. Paris, Bibliothèque nationale de France. MS Latin 14339, fol. 251v. With permission of Bibliothèque nationale de France.

established by figures wearing lower-class garb: the stereotypical herdsman at left shouldering a pole with dangling basket, leading an ox by a rope tied to its horns, and accompanied by a goat and a flock of sheep; at right rears a horse with a bare-legged rider. The herdsman's feet are bare and he is clad in a short, ragged muddy-colored garment; the barefoot rider is also dressed in a short robe that exposes his knees and wears a rustic broad-brimmed hat. In contrast, at center stand the aristocratic landowner with sword and fur-lined cape and his legal agent, both shod and wearing long robes, haggling over the price of the livestock.

Urban scenes are generally populated by long-robed figures, whose identity is most often established by the context or by an object particular to their profession. Questions of legacies and inheritance are represented by compositions featuring a dying man dictating his last will and testament, along with other individuals who participate in the act of dying. In a Bolognese fourteenth-century miniature at the Vatican Library[15] the dying man lies in his bed at center, comforted by his dutiful wife standing at the head of the bed, and attended by a kneeling tonsured cleric who will celebrate mass at the proper moment. The customary scribe sits at foreground with scroll on his lap, annotating the dictated bequests. A doctor and his assistant stand at far right, the eminent status of the doctor signaled by his fur collar and the fur-lined sleeveless cloak. Here, however, costume-as-signifier is outweighed by the doctor's most telling attribute—the urine flask he holds up to the light.

Criminals, malefactors, and those associated with them—hangman and executioners—are specially differentiated in the context of crime and its punishment. First of all, criminality was visually equated with the lower classes of society, including country folk, street peddlers, tradesmen, and manual laborers, distinguished in medieval representations by more modest or abbreviated clothing such as that worn by the bare-legged rural folk in Figure 8.4. In general, short robes and varying amounts of bare skin, often coupled with aggressive gestures, designated an individual of inferior social class or of questionable merit. In illustrations to Justinian's *Codex* and *Digest,* the most commonly portrayed criminal is a thief, usually short in stature, occasionally given coarse features such as a pouting lip or pug nose, and sometimes depicted bald. A smooth hairless head could have pejorative implications: hair was often shorn as a defamatory punishment, marking the individual for public derision. In typical miniatures the thief is brought before a judge for sentencing, many times stripped to his undergarments, with his arms usually bound at the wrist.[16] He may be clothed in a short garment in contrast to the long robes of lawyers and jurists, and he is sometimes crippled or missing a limb,[17] the sign of a recidivist. In a fourteenth-century illustration to Justinian's

Codex, Book 6,[18] (Fig. 8.3) a thief, clutching the book he has stolen, is being pummeled by his outraged captor. His strongly negative visual profile is enhanced by the exposure of bodily features normally concealed. He is bald and naked except for a strategically draped swag of cloth, but is also portrayed with sprouts of underarm and chest hairs. In Renaissance painting this last detail would represent a touch of verisimilitude, but in the 1340s the portrayal of body hair other than on the head and face expresses degeneracy and social undesirability.

The individuals responsible for discipline and punishment, hangmen and executioners, are sometimes distinguished by special headgear: short red pointed hoods or head-enveloping versions cut with eyeholes.[19] The figures who wear these hoods either function as police escort to the criminal being brought before a judge, or actually administer punitive measures such as blinding, branding, or amputation of the hands or feet. The masking hood was provided for people exercising this profession to hide their identity, for in most places the executioner was shunned from normal society and his touch was considered to transmit infamy.[20] In other instances infamous character was transmitted by facial appearance or expression, and in two miniatures from a manuscript in Siena[21] the illuminator characterized the hanged criminal and the executioner with the same grotesque, grimacing visage.

Articles of clothing or accessory items may be charged with their own meanings, and we now turn to examine costume as object. In certain situations a garment itself—extended, accepted, and being vested, or rejected, removed and thrown away—becomes a signifier for transition: of profession, level of responsibility, or social status. The first comprehensive compendium of canon law regulations and procedures, Gratian's mid-twelfth-century *Decretum,* discusses in Causa 20 whether a child should be obliged to enter the priesthood if there is no definite sign that this is his true vocation. It uses as a case study the story of two young boys who were taken to a monastery by their parents, probably second sons, to enter as novices. Only one of them was willing, however, and the other rebelled against taking the cloth, probably wailing in despair "Daddy, I don't want to be a monk!" How should an artist portray this real-life situation, discussed in a manual of canon law?[22]

Early illustrations depicted one boy standing or kneeling before the receiving abbot, the reluctant one looking back appealingly at his parents.[23] Others pictured the compliant youngster stretching out his arms to the receiving brothers, while the rebellious child tried to flee, darting away from restraining parental arms.[24] In the late thirteenth century, however, artists hit upon the idea of using the monastic robe itself to symbolize the acceptance or rejection of the religious life. In numerous manuscripts we see the unwilling youngster dashing away in his lay clothing, while his companion, sometimes already with mini-tonsure, often stripped to the waist, kneels to be vested in the black robe.[25] A more narrative version dating to the 1340s[26] transposes the scene to a nunnery, where at center the willing little girl, hair already shorn, is about to be vested, and at right another nun extends a robe to a second candidate whose hair is about to be cut by her mother. At far left rejection is communicated by a discarded white robe, flung to the ground in a heap by the pint-sized fugitive who looks back briefly in her flight.

In another *Decretum* case study for Causa 17 the robe takes on the status of a *desired* object, when an ailing priest—in the earliest compositions pictured leaning on a crutch, in later versions depicted on his sickbed—vows to give up his church and benefice and become a monk. In these miniatures surrounding members of the monastic community hold up or extend to the priest the monastic robes.[27] This ceremonial gesture marks in essence the passage from a more worldly existence to the contemplative life, affected, we may suppose, by the cleric's hope that his piety would help him regain his health. Conversely, in an illustration to a *Decretales* passage,[28] we see the removal of the desired object as a cleric, having renounced the church, is stripped of his habit.

On the secular side, articles of clothing may be used to represent legal decisions or regulations. Various books of Justinian's *Digest* deal with issues of marriage and betrothal, dowry laws, and the disposition of dotal property when a marriage is ended. Book 24 opens with a discussion of the need to regulate gifts between husband and wife, stating: "As a matter of custom, we hold that gifts between husband and wife are not valid. This rule is upheld to prevent people from impoverishing themselves through mutual affection by means of gifts which are not reasonable, but beyond their means."[29]

Most miniatures for this textual location feature a couple standing face to face, the man stripped to his *braies,* chest bared, his other garments held by the woman. The most well-known, a splendid version by the Bolognese Master of 1328[30] has usually been interpreted as an intimate domestic scene depicting a wife helping her man to dress as the children look on,[31] since it is staged within a bedchamber, where a woman glances tenderly at two toddlers and an infant in a cradle as she appears to hand some garments to the semi-nude male figure before her. This composition had been in force at least by the 1280s or 90s, as evidenced by the two-part miniature in Figure 8.5, the earliest example I have seen.[32] Here we see another side to the story. In the left-hand section the woman appears to be pulling off the man's robe, witnessed by a child peeping around the dividing column; at right the man is virtually naked with only a strip of cloth pulled

Figure 8.5 Justinian, *Digestum vetus*: Book 24, 1280–90. Oxford, Bodleian Library. MS. Canon. Misc.493, fol. 420. With permission of the Bodleian Library, University of Oxford.

over his genitals, his head resting on his hand in a posture of lament. A figure standing before him seems to commiserate with him. In a later fourteenth-century version,[33] the disrobing transaction takes place in the
presence of two witnesses, members of the woman's entourage. Despite the
genre touches in the Turin miniature, the composition for this book was
obviously not originally intended to be a representation of domestic inti-
macy. The key word in the opening passage to Book 24 is the verb, *impov-
erish,* and the illustrations refer to the phrase "people impoverishing
themselves through mutual affection by means of gifts which are beyond
their means." In reality, these miniatures depict the groom literally making
a gift to the bride of "the shirt off his back," in effect, providing too large
a wedding gift or *Morgengabe* for his financial resources. A contemporary
fourteenth-century miniature[34] more accurately expresses the juridical
character of this transaction: within a courtroom the bridegroom strips
himself before a judge, while the bride-to-be takes his clothing piece by
piece, and a notary records these gifts in an official ledger. This subject
would have been especially relevant to the inhabitants of Bologna, where
communal decrees in the thirteenth and fourteenth centuries regulated the
number of rings a man could give to his bride, the size of the bridal party,
and the percentage of the bride's dowry that the groom was expected to
contribute on his part.[35]

As a concluding point, we will look at how the presence, absence, or
physical aspect of an article of clothing may express a legal condition. In
canon law manuscripts, some of the most dramatic illustrations were made
for cases concerning courtship, marriage, and sexual relations between men
and women. The illuminator of a *Decretum Gratiani* at the Fitzwilliam Museum (MS 262) provided very explicit representations of sexual activity in
his compositions for Causae 33 and 36, the first concerning a case of adultery and the second a case of rape. Consensual and forced sexual congress
is visually distinguished in the two miniatures by subtle differences in gesture and dress. The case study for Causa 33 discusses a husband who had
become temporarily impotent. As a result, his wife denounced this situation in a church court, and subsequently committed adultery with and
married another man. When the first husband recovered his potency, however, the wife was obliged to separate from the second illegal husband and
rejoin her first. This is pictured in a narrative sequence in four compart-
ments,[36] where in the upper right compartment the adulterous coitus takes
place in a tangle of clothing and limbs. The vigorous activity exposes the
woman's feet, shod in elegant black slippers with pointed toes, and her slim
legs clad in knee-high stockings, secured by garters. Although there is a
glimpse of bare knees and thighs, decorum is preserved because most of
her body is covered, and the docile manner with which her arms enfold

the male figure above her suggest that she submits willingly to him. Causa 36 involves the question of whether and how, under church law, a couple that has already had sexual relations may be joined in marriage. For this theme the Fitzwilliam artist[37] illustrates a situation known today as "date rape": a young man invites a young woman to dinner, induces her to drink, and then takes advantage of her intoxication to debauch her. Once again the artist presents the story in a narrative cycle: in the first episode the young man encounters the young woman accompanied by her father; the second pictures the seduction at the dinner table; and at lower left we see the young woman's rape. In this scene aggression and force are conveyed by various details: the young man mounts her brusquely with penis exposed, and his right hand extends to fondle the woman's chin in a long-recognized gesture of male possession.[38] But it is the young woman's disarray that constitutes final proof of violation: here shoes and stockings have been removed, her legs are entirely bare, and her dress is roughly pushed up to expose her vagina. On such visual evidence, many a woman's reputation could be sustained or destroyed.

Notes

1. New York, Pierpont Morgan Library, MS M.716.1, dating to the 1330s; the miniature illustrates the Prologue to the *Decretales* of Gregory IX.
2. Weimar, Thüringische Landsbibliothek, MS Fol. max. 10, fol. 6, illustrated in Alessandro Conti, *La Miniatura Bolognese,* Bologna, Edizioni Alfa, 1981, color plate XXX.
3. Siena, Biblioteca Communale degli Intronati, MS K.I.10; illustrated in Grazia Vailati von Schoenburg Waldenburg, "La miniatura nei manoscritti universitari giuridici e filosofici conservati a Siena," in *Lo Studio e i testi: Il libro universitario a Siena (secoli XII–XVII),* ed. Mario Ascheri (Siena: Comune di Siena, Biblioteca Comunale degli Intronati, 1996), 79–144, figs. 13, 14, 17, 18.
4. For marriage customs in Tuscany and northern Italy ca. 1300–1500, see especially "Zacharias, or the Ousted Father: Nuptial Rites in Tuscany between Giotto and the Council of Trent," in Christiane Klapisch-Zuber, *Women, Family, and Ritual in Renaissance Italy* (Chicago: The University of Chicago Press, 1985), 178–212; for a detailed discussion of European marriage miniatures and their iconography, see the recent dissertation by Kathleen Nieuwenhuisen, *Het Jawoord in Beeld: Huwelijksafbeeldingen in middeleeuwse handschriften (1250–1400) van het Liber Extra* (Ph.D. Dissertation, Academisch Proefschrift, Vrije Universiteit te Amsterdam, 24 November 2000).
5. Cambridge, Fitzwilliam Museum, MS McClean 136, fol. 188.
6. Causa XXX, *Quaestio* V, can. 3.
7. Vatican City, Biblioteca Apostolica Vaticana, MS Vat. lat. 1389, fol. 241.

8. For legal dress in general, see W. N. Hargreaves-Mawdsley, *A History of Legal Dress in Europe Until the End of the Eighteenth Century* (Oxford: Clarendon Press, 1963).

9. Vatican City, Biblioteca Apostolica Vaticana, MS Urb. lat. 164, fol. 1.

10. Among them Madrid, Biblioteca Nacional, MS 1548, fol. 1 and Cesena, Biblioteca Malatestiana, MS S.IV.1, fol. 1.

11. Munich, Bayerische Staatsbibliothek, MS Clm 21, fol. 175v.

12. Jonathan Alexander, "*Labeur* and *Paresse:* Ideological Representations of Medieval Peasant Labor," *Art Bulletin* v. LXXII (1990), 438.

13. See especially the miniature by Niccolò da Bologna, Cambridge, Fitzwilliam Museum, MS 331.

14. From Justinian's *Digestum vetus,* Book 18: *De contrahenda emptione et de pactis inter emptorem et uenditorem compositis et quae res uenire non possunt* (Conclusion of the contract of purchase, special terms agreed between the vendor and purchaser, and things that cannot be sold); Paris, Bibliothèque nationale de France, MS Latin 14339, fol. 251v.

15. Vatican City, Biblioteca Apostolica Vaticana, MS Vat lat. 1436, fol. 40, illustrated in Conti (as in note 2), fig. 284.

16. In thirteenth-century manuscripts, both manumittees and thieves were represented in this manner. Some examples are Oxford, Bodleian Library, MS Canon misc. 495, fols. 171 and 216v; Toledo, Archivo y Biblioteca Capitulares, MS 32–15, fol.185; Milan, Biblioteca Ambrosiana, MS D.533. Inf., fol. 175. For some iconographical interpretations of hair and headgear see François Garnier, *Le langage de l'image au moyen âge: Signification et symbolique,* v. II: Grammaire des gestes (Paris: Le Léopard d'Or, 1982, especially 78–83); Ruth Mellinkoff, "Demonic Winged Headgear," *Viator* 16 (1985), 367–381ff; and her *Outcasts: Signs of Otherness in Northern European Art of the Late Middle Ages* (Berkeley: Los Angeles, and Oxford: University of California Press, 1993), for the iconography of malefactors in general.

17. As in the miniature for Book 6 in Vienna, Österreichische Nationalbibliothek, MS Cod. 2052, fol. 183.

18. Paris, Bibliothèque nationale de France, MS Latin 14339 fol. 203.

19. Examples are found in two miniatures illustrating Justinian's *Digest:* Vienna, Österreichische Nationalbibliothek, MS Cod. 2252, fol. 147 (for Book 48), and Oxford, Bodleian Library, MS Canon misc. 493, fol. 21v (for Book 2).

20. See Pieter Spierenburg, *The Spectacle of Suffering* (Cambridge: Cambridge University Press, 1984), 13–20.

21, Justinian, *Digestum novum,* Siena, Biblioteca Communale degli Intronati, MS I.IV.5 fol. 173 and fol. 197v.

22. See the numerous illustrations ranging from the late twelfth to the fifteenth century in Anthony Melnikas, *The Corpus of the Miniatures in the Manuscripts of the Decretum Gratiani* [Studia Gratiana XVI-XVIII] (Rome: 1975), vol. 2, 631–54.

23. As, for example, in Siena, Biblioteca Communale degli Intronati, MS K.I.3, fol. 216v, and Cambridge, Corpus Christi College, MS 10, fol. 212, the latter illustrated in Melnikas (as in note 22), Causa XX, fig. 12.

24. Such as in Vatican City, Biblioteca Apostolica Vaticana, MS Vat. lat. 1375, fol. 201v or Vatican City, Archivio della Basilica di S. Pietro MS A.25, fol. 193v, both illustrated in Melnikas (as in note 22), Causa XX, figs. 31, 42.

25. Illustrated in Melnikas (as in note 22), Causa XX, figs. 33–44.

26. Vatican City, Biblioteca Apostolica Vaticana, MS Vat. lat. 1366, fol. 198, illustrated in Melnikas (as in note 22), Causa XX, color plate V and in Conti, (as in note 2), fig. 268.

27. See especially Troyes, Bibliothèque Municipale, MS 103, fol. 149 and Paris, Bibliothèque Nationale, MS Latin 3898, fol. 215v, illustrated in Melnikas (as in note 22), Causa XVII, figs. 11and 34.

28. Chantilly, Musée Condé, MS XVIII.E.1, fol. 75.

29. *Moribus apud nos receptum est, ne inter uirum et uxorem donationes ualerant. Hoc autem receptum est, ne mutuo amore inuicem spoliarentur donationibus non temperantes, sed profusa erga se facilite.*

30. Turin, Biblioteca Nazionale Universitaria, MS E.I.1, fol. 310, illustrated in Conti (as in note 2), fig. 245.

31. "while helping her husband dress, the wife casts a concerned glance towards the three children behind her." Quoted from Patrick M. de Winter, "Bolognese Miniatures at the Cleveland Museum," *Bulletin of the Cleveland Museum of Art,* 70 (October 1983): 328.

32. Oxford, Bodleian Library, MS Canon misc. 493, fol. 420.

33. Vatican City, Biblioteca Apostolica Vaticana, MS Vat. lat. 1411, fol. 365.

34. Paris, Bibliothèque nationale de France, MS Latin 14339, fol. 321.

35. See Lodovico Frati, *La vita privata di Bologna dal secolo XIII al XVII* (Bologna: Zanichelli, 1900 [reprint Bologna: Arnaldo Forni Editore, 1986]), 49–51.

36. Cambridge, Fitzwilliam Museum, MS 262, fol. 86v; illustrated in Melnikas (as in note 22), Causa XXXIII, fig. 33.

37. Cambridge, Fitzwilliam Museum, MS 262, fol. 137; illustrated in Melnikas (as in note 22), Causa XXXVI, fig. 35.

38. See Diane Wolfthal, "A 'Hue and a Cry': Medieval Rape Imagery and its Transformation," *Art Bulletin* (March 1993): 39–64 and also chapter four in her *Images of Rape: The "Heroic" Tradition and its Alternatives* (Cambridge: Cambridge University Press, 1999), 99–126.

PART THREE

THE LATE MIDDLE AGES

CHAPTER 9

FROM BATTLEFIELD TO COURT:
THE INVENTION OF FASHION
IN THE FOURTEENTH CENTURY

Odile Blanc

The fundamental transformations that influenced men's way of dress in the fourteenth century established decisively the differences between the sexes. At the same time they inaugurated, in the Western history of fashion, modern ways of dressing by adopting short, fitted, and tailored[1] elements as new criteria for elegance. The decades before and after 1400 are distinguished by such a diverse vestimentary landscape that it evoked a "Babel of costumes" for the great historian Jules Michelet. The period could equally be defined as "the age of the *pourpoint*," since at this time the military garment served as an emblem of the martial function as well as a courtly mode.

Enormis Novitas

From the middle of the fourteenth century, in Italy as well as in France, England, Germany, and Bohemia, numerous chroniclers note that among their contemporaries there was a sudden transformation in manners of dress. Most of these texts were written well after the events they describe took place, and it is important to emphasize that from that time onward they are thus integrated into a narrative whose aim is essentially moralizing. By the yardstick of tradition, history had as its charge to record any novelty that appeared as a perilous disturbance. The century in which the chroniclers wrote was marked by the multiplication of armed conflicts, born out of a succession of quarrels between England and France. This period is better known under the name of "The Hundred Years' War," an age of rivalries between local powers, especially in cities where local lordship had been well established. This was also the time of famine, poverty, persecutions of heretics and Jews, and epidemics. In these texts, mentions of the Black Death in the middle of the century were especially frequent. The authors passed judgment on the changes in dress through observations of contemporaries, worthy of mention in their writings. The new styles are described as important mutations, revelatory of the calamity of the time, and sometimes as premonitory signs of the approaching end.

Thus, in Rome, the anonymous biographer of Cola di Rienzo, often called the *Anonimo romano*,[2] discussed the vestimentary transformations of the years 1339–1343. He drew attention to what he considered ill omens: the apparition of a comet; a famine due to bad weather; and the battle of Parabiago (1339) during the course of which members of a prestigious family, the Visconti, confronted one another. In 1340, Jean de Venette, a Carmelite friar at the convent at Place Maubert in Paris, probably was a witness to the events he described as he continued the historical account begun by his predecessors.[3] The year 1340 was full of calamities for the kingdom of France: vestimentary changes again took place after the apparition of a comet and the victorious expedition of the English king, Edward III, at Sluys in Flanders. Gilles le Muisit, the abbot of the Benedictine monastery of Saint Martin of Tournai, dictated his *Annales*[4] between 1350 and 1353, and reminisced about the year 1349 as a period of great disorder. The Plague struck the region of Tournai in this year, taking the life of the bishop. Then there were several processions by the Flagellants, disseminating fear in their wake. Finally, new dress outfits made their appearance, one more disgraceful than the other. In England, the *Chronicles of Westminster*[5] was composed as the official historiography whose equivalent on the other side of the Channel was the *Grandes Chroniques de France*. In the year 1365, one of its authors, John of Reading, reported on the vestimentary changes among his English contemporaries in a similar, apocalyptic context.

All of these chroniclers express shock that one could no longer see a difference between the nobles and their servants, the clerics from the lay people, and the men-at-arms from the civilians. The men dressed in short garments were as affected in their outfits as the women were, thus bringing offense to their manly customs. And the elderly, who ought to set a better example, did not hesitate to adopt the new fashions. This sort of complaint conforms to a long tradition of the condemnation of luxury, voicing an opposition against the corruption of the morality of the day, and, due to an excess of refinement, the abandonment of the ascetic life of the ancients. After the great French defeats by the English (Crécy in 1346, Poitiers in 1356), the official historiography thus blamed the national calamity on such new mores, and castigated the adoption of the foreign fashions. In the case of Florence, recalled by the frequently cited historian Giovanni Villani,[6] the change in dress coincided with the rise to power of Gauthier de Brienne, nephew of Robert of Naples. He was a knight-mercenary who served in the army of the king of France, hence the French origin attributed to the new clothing styles in Italy. His entry into the city in1342 was a military one, aimed to establish his authority. The new mode of dress also became a symbol for the illegitimate government of Florence that had appointed a foreigner as head of the city-state. Only the author of the chronicle of Limbourg,[7] a cleric in the diocese of Mainz who began to write in 1377 on the events that had taken place between 1335 and 1398, interpreted these changes as a sign of renewal after the Black Death. After this great mortality, he wrote, people set out again to live, to become joyous, and men had new kinds of garments made.

Illuminated Manuscripts as Fashion Illustrations

Except for the isolated voice of the cleric of Mainz, the words of the chroniclers were those of inflexible moralists who had nothing positive to say on the point of view of the new male fashions. It is here that the sur-viving images turn out to be essential. The miniatures of the prestigious manuscript of the *Grandes Chroniques de France,*[8] made for Charles V around 1375–1379, appear as veritable odes to the novel ways of dressing, at the same time as their texts register disapproval of them. The new sil-houettes seen in the different categories of military men, servants, and members of the aristocracy can be found in this work, and in many other illustrated books of the same period. In truth, they are so widely prevalent in the iconographic representations that they appear to correspond to common usage, and not to the more or less marginal extravagance em-phasized in the Limbourg chronicle. This "current usage" was nothing but the construction of fashion itself, prescribed by the models portrayed, and as they were expressed in a given moment among the individuals situated within it, conforming to it or deviating from it—all long before the term "fashion" makes its entrance into the dictionaries.

For a long time the miniatures in manuscripts have served as straight-forward illustrations to the writings, relating contemporary fashions. At the same time, they have often been suspected of embellishing reality, since only the practices or fantasies of the aristocratic milieu are reflected. The illustrated books were in fact commissioned by and executed exclusively for the wealthiest categories of society, those who actually wore the fash-ion garments that were denounced by the moralists. These exceptional, luxurious documents were the products of an elite representing itself the way it would like to appear, and the images in them constitute an essential source toward understanding the vestimentary practices of the period. The diversity of dress presenting itself in these books is entirely in the service of the imagination of the aristocratic culture. When the figure of the peasant in rags appeared, as in the calendar pages of the *Très Riches Heures* of the Duke of Berry, for example,[9] it is indeed both an aristocratic, and a belittling representation of a man of the people that we are given to see. It is not possible to say to what degree this figure is devoid of realism and has emerged from pure fantasy. The image exists, to be sure, but the man is represented as the aristocratic sensibility wants him to appear. Likewise, the elegant silhouettes are not "immediate," in the photographic sense of the term, representations of medieval courtly life or everyday life.

As the fashion photography trade of our day subjects actual, real garments to a particular, artificial staging, so did the medieval illuminator bring into view true enough pieces of clothing. Then as now our patterns of association emerge from a veritable *bricolage* that expresses the sensibility of the time, and defines the actuality of the vestimentary practice, as Roland Barthes[10] has so effectively shown. In this sense, and in the absence of a specific "fashion discourse," one can compare an illuminated manuscript to a contemporary fashion magazine. Just like it, the miniature proclaims the new model, for which it is at the same time the producer and the broadcaster. While the medieval authors had difficulty breaking away from the traditional Christian morality in which fashionable appearance is the symbol of worldly vanities, the painters commissioned by the princes had the task of representing differences through degrees of opulence in dress. At a time when courtly life was in the process of development, the appearance of one's clothing was the marker of power and splendor.

The vestimentary effervescence around 1400 coincides with a "golden age of illumination."[11] John of Berry, brother of Charles V, certainly employed the best artists of the time. His library contained about three hundred books, an enormous number in its time, and several of them rank among the most remarkable examples of illuminated manuscripts. The seven books of hours that he commissioned have been the focus of attention of many art historians. But John of Berry was equally keen on collecting chronicles, chivalric literature, and translations of works from antiquity, which were interests he shared with Charles V. Philip the Bold, duke of Burgundy, was a cousin to John of Berry, and prided himself of an almost equally prestigious library, although of lesser importance. His son John the Fearless distinguished himself less by his taste for beautiful books than for his political contests, while the library of his rival, Louis of Orléans, contained many illuminated and richly bound volumes. In comparison to the extraordinary expansion of the libraries of the endowed princes, the royal book collection appeared a poor parent. In view of the endeavors of his predecessors John the Good and Charles V—particularly the efforts of Charles V to develop translations, copy classics, and encourage artistic projects—the patronage of Charles VI appears lusterless, an evident sign of the weakening of royal power.

The striking contrast between the political disorder and the rise of the artistic production has not escaped the notice of the historians. After Michelet, Johan Huizinga put his stamp on generations of readers by the subtle term he coined, "the autumn of the Middle Ages," full of contrasts, spreading a "mixed smell of blood and roses."[12] Around 1410, John of Berry commissioned the sumptuous prayer book previously mentioned, the *Très Riches Heures*. At this time, Paris was the scene of bloody encounters between the Armagnacs—partisans of duke Louis of Orléans, who was assassinated in 1407 by the order of his cousin John the Fearless—and the Burgundians grouped behind the latter. Put to the test by war and by the political and economic chaos that accompanied it, the aristocratic society evolved within the confines of the closed world of the court, and made itself visible, idealized, in the illuminated manuscripts. It is here that a new fashion model was elaborated, that of the courtier, no longer to be understood solely by feats of arms, but in the representation of self, manifested by his mastery of social codes and his belonging to the courtly elite.

Le corps guerrier—The Martial Body: A Model

In 1342, Giovanni Villani, the previously cited Florentine historian, described the young men dressed in *cottes* or *gonelles* that were so short and tight that they could not be put on without help. They girdled themselves with straps, resembling horses' saddle girths, with enormous buckles, points and voluminous purses "*à l'allemand*" suspended over their bellies. They arranged their hoods in the way of monks, that is to say they wore them as cowls, hiding most of their faces, with a cape that descended to the belt or beyond it. The edges of this cape were decorated and slashed, and the *cornette*, the tip of the hood at the top of the head, became elongated to reach the floor when it wasn't coiled around the head. These hooded men wore long beards that gave them the air of fierce warriors. The knights wore a *surcotte* or a tight fitting and belted *ganache* with elongated sleeves, *manicottoli,* that swept the ground. These were the new silhouettes. The former was adopted by a class of young men whose status was not yet defined, and who searched for fortune in war or at tournaments.[13] The latter was reserved for the superior category of knights, constituting the elite of the armed forces.

For a long time, the *cotte* designated the garment worn directly over the shift, and under another garment called the *surcot,* in its turn worn under a cape, a mantle, or a cloak. This ensemble constitutes the "*robe*," making up the wardrobe components of a nobleman, of which the layers of garments (and by consequence their number) indicate the rank and wealth of the wearer. In representations, the mantle, for example, invariably distinguishes the prince, and lay or ecclesiastical dignitaries in the exercises of their duties. The manner in which the outer garments allow the lower garments to show is essential to every demonstration of appearance; this is a device that does not disappear, but is retained with the new garments. The knights described by Villani also wear *surcots* whose trailing sleeves reach the ground while they also reveal the clothing worn beneath, as can be seen in many images. In a French manuscript of the middle of the fourteenth century, now in the Vatican Library,[14] the princes received by Agamemnon wear garments that reach only to the calf. Although they were ample below, they were held in without pleats at the chest level, indicating a fitted cut. The sleeves were cut at the elbows and fell in sweeping panels although in still modest lengths. These were the *coudières* described by numerous authors, named *manicottoli* by Villani. The sleeves, so form-fitting they might have been actually stitched to the skin, opened at the level of the elbow not to reveal flesh, but the sleeve of a garment underneath. In the same way the lower edges of the garment included long slits in several places, revealing not only its lining but also, probably, the *cotte* to be found underneath. In this way, no matter what the chroniclers said, the traditional *cotte/surcot* system prevailed, at least in this social category. But the visible superposition of garments found itself modified by the tightening of the upper parts of the clothing.

Another example of these transformations can be found in the papal palace in Avignon.[15] In this magnificent princely residence where the best (primarily Italian) artists of the time worked, one can still admire two young and elegant nobles preparing themselves for a hunt with falcons on the walls of the *Chambre du cerf*, painted in 1343. (Figure 9.1.) They are dressed in precious garments reaching to the calves, with the fabric held in at the upper chest, then more ample below, and at the bottom forming a sort of gathered skirt, indicating a seam at the junction of the two parts. The waist is low and finely marked by a belt from which a dagger is suspended, a practice noted by the chroniclers with some emphasis. The sleeves have the *coudières* characteristic of the time, opening to reveal the narrow sleeves in contrasting color worn underneath. Finally, one of the young aristocrats has put on an equally short mantle over his elegant *surcot*. Similar examples represent the wearing of a large leather belt, adorned with a voluminous purse covering the abdomen. This is seen in a manuscript made in Toulouse around 1350[16] that once belonged to the count of Foix, Gaston Phébus, a great lord and commissioner of illuminated manuscripts. Fitted down to the hips, the garment then widens, and ends just below the knees. The long *coudières* reveal the tight, completely buttoned sleeves of the undergarment, and the figure is wrapped in a hooded cape, the base of which covers the top of the body to the chest. This is an example of the over-garment with a long *cornette* and ample cape so disparaged by the chroniclers. A psalter that once belonged to Bonne of Luxembourg,[17] now in the Metropolitan Museum of Art in New York, also bears witness to the shorter and more fitted clothing, accessorized by *coudières,* hooded capes, and leather purses.

In a text well known to costume historians, the author of the *Grandes Chroniques de France* attributed the rout of the French troops to the vestimentary excesses of the knights, first at Crécy in 1346, then at Poitiers ten years later. However, not a single garment is named. Some, said he, had such short clothing that their rumps were barely covered, and so tight that they needed help to undress. Others wore garments like women's, gathered at the small of the back, with *coudière* sleeves, and capes with slashed edges. This latter silhouette appears to be similar to those just discussed. The former, however, is evidently a much shorter article of clothing, necessitating an opening all along the front due to its tight fit, a shape that will be imposed during the sixties and seventies of the fourteenth century. It seems that previously the two types of clothing had coexisted.[18]

The author of the Limbourg chronicle describes a garment that appeared, according to his memory, after the Black Death at the middle of the century. This novel dress item, very short and fitted, is not cut below the small of the back, and is made of many pieces of fabric (*geren*). Without a doubt, we must understand that this garment was subjected to the new techniques of tailoring that multiplied the number of seams in order to do away with pleats that had become awkward. These cleverly assembled fabric pieces obviously played the role of darts used in today's tailoring. In England, John of Reading described the fashions of around 1365 and emphasized that the close-fitting men's clothing had been stitched from many parts, "*cousus de toutes parts.*" An illuminated manuscript of the works of Guillaume de Machaut,[19] executed around 1350–1355, represents a number of slender, masculine silhouettes in sheath-like garments, short and fitted in their entirety, slipped on over the head, it appears, like a pullover of today. Some are buttoned all along the front, others appear to be composed of horizontal bands that evoke the characteristic quilting of military garments or *pourpoints.*[20] These silhouettes hardly resemble those previously cited, and therefore refer without a doubt to another way of dressing, while still being associated with hoods having wide cowls.

In his chronicle, the Mainz cleric remarks not without astonishment that during the period that followed the Black Death, lords, travelers,

Figure 9.1 *Preparing for the Hunt.* Wall painting, Chambre du cerf, 1343. Avignon, Papal Palace. By permission.

knights, and their servants all wore armor, and that the *pourpoints* were, according to them, reinforced with iron plates. And in 1358, the Parisian tailors demanded the rights to make them, an activity until then reserved only to the *pourpointers*. The tailors argued that this garment had become so widespread that to meet the demand, it would not be excessive to establish two guilds to make them.[21] In the year 1367 of his chronicle, the

Prague canon Benesch de Weitmühl[22] mentioned a novelty that one does not find among any of the previously cited authors. Then, he wrote, his contemporaries had made for themselves short and fitted garments that featured a padded fullness at the breast like that of the bosom of a woman, and these were so tight that they look like the hunting dogs called grey-hounds. As an echo of the history written by the cleric of Mainz, this text also brings attention to the evolution in the form as well as in the way of wearing the *pourpoint,* constituting precisely the vestimentary novelty of the fourteenth century.

The spread of armed conflict over all of medieval Europe caused men more and more to adopt defensive garments hitherto reserved for martial activities, and of which the style of the garment worn uppermost was modified. This S-shaped silhouette was fundamentally different from that of the Avignon *Palais des papes,* and that of men-at-arms, knights, and servants that populate the illuminations of the *Grandes Chroniques de France,* and several other manuscripts of the same period. The garment worn by Jean de Vaudetar has become a classic example of the fashions of the seventies of the fourteenth century, seen in the celebrated dedication scene[23] in which the counselor offers the Bible to Charles V that he commissioned. (Figure 9.2.) The garment is fitted very close to the body, the waist is pinched, and the bosom juts forward in the strange bulge that did not fail to astonish the chroniclers. A richly bejeweled girdle rests on his hips, and from it is suspended a dagger, thus displacing the body's apparent center of gravity.

The Musée des tissus in Lyon has a garment exactly like the one worn by the counselor. (Figure 9.3.) Executed in a precious, white silk brocaded in gold, it is composed of 32 pieces of cloth, confirming the skills of the tailors of the period. Closed in the front by 32 buttons, the garment has narrow sleeves also set with buttons, and of a type called "*grandes assiettes,*" the flat elements that define the armscyes of the sleeves. The extent of the "*assiettes,*" reaching almost to the center of the body, assures a most precise fit. Similarly, the triangular pieces that are regularly disposed around the armscye and recall the "pieces" mentioned by the Mainz cleric, make any pleats unnecessary in this part of the body. This garment is entirely padded with cotton batting, held in place by horizontal stitches that are not visible on the reverse. Finally, the prominent chest implies that another, fitted garment must have been worn underneath to support this effect. The Lyon garment, then, was probably worn over a *pourpoint,* which gave it its characteristic form. One can compare it to the *jaques* and *jaquettes* listed in the inventories of the time of Charles VI, designating luxurious items in princely wardrobes. The traditional association between *cotte* and *surcot,* still present around 1340, has now completely disappeared.

Figure 9.2 *Jean de Vaudetar offers his Bible to Charles V.* The Hague, Museum van het Boek/Museum Meermanno-Westreenianum. MS. 10 B 23, Fol. 2. By permission.

Around 1400 the armscyes made with *"grandes assiettes,"* until then rarely seen in surviving images, are superseded by the ample, hanging sleeves that doubtless belong to the *pourpoint* properly speaking. They are associated with other, very short and fitted garments, but also with a new mode of dress once again with a lower part ample and long, the *houpplande.*

The Construction of Appearances

Under the rule of Charles VI, the short garment that established itself during the course of the fourteenth century was in its turn challenged by a new, longer item of clothing. Inventories continue to mention sumptuous *pourpoints, jaques,* and *jaquettes,* but the images show them reserved as undergarments, covered by the *houpplande,* or seen in special circumstances such as martial exercises. After the duress of the many wars, the *pourpoint* again became what it once was: a military garment. The short garment born out of it was for the most part invisible. Nevertheless, the promotion of the short outfit ushered in a new vestimentary era during which the tailored triumphs over the draped, and in which a new way of clothing the two genders emerged into the light of day, henceforth destined to fashion the bodies distinctly and separately.

For a long time historians of costume have seen in the transformations in dress at the end of the Middle Ages the heralding of the rediscovery of the human body that, according to them, characterizes the Renaissance. Certainly, the short garment was worn close to the chest and exhibited the formerly concealed legs of the male body. But this leveling of the anatomy was not preceded by a denudation of the body as a way to body consciousness. Far from being a docile glove fitting the morphological form, the short garment was a rigid envelope that constricted the body into narrow confines and imposed a curious posture. The cut of the sleeves of the garment preserved in Lyon, for example, adjusts with difficulty, a priori, to an extended arm position. The garment rather appears to approximate the shape of the body and necessitate the use of a corset to achieve it, while its silhouette strangely evokes the sinuous female fashions of the end of the nineteenth century.

Far from being a return to the natural body, the new, fourteenth-century clothing styles seemed ceaselessly to produce different types, resulting in a range of garments. The extremities from now on were seen as distinct from the rest of the body, and in certain respects they appear to recall essential morphological traits. Thus, the buttoning of the garments from top to bottom superimposes itself on the vertical axis of the body around which symmetries—or asymmetries, as in the case of the *mi-parti,* or bi-color dress—are ranged. The girdle marks the point of equilibrium, displaced to the hips. The

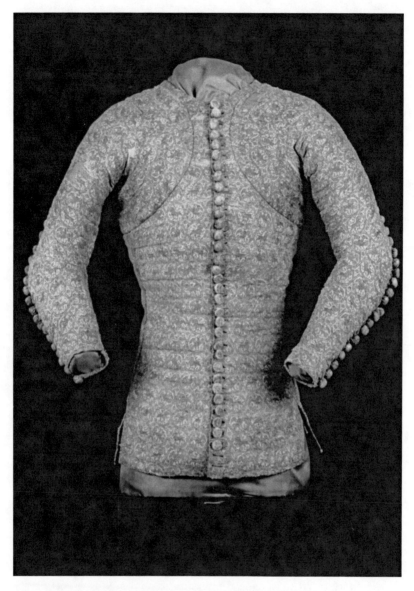

Figure 9.3   Pourpoint of Charles of Blois, c.1364. Lyon, Musée des Tissus. Photo credit: Stephan Guillermond. By permission.

coudières separate the forearm from the lower arm. All these elements function as markers of the anatomical body, underscoring points of articulation. This dynamic does not disappear with the return of the longer garments, to which a miniature bears witness in a Boccacio[24] manuscript, exemplary of the fashions at the turn of the fifteenth century. In this image, representing Venus and her admirers, a figure in the foreground wears a very short garment, cinched at the waist by a voluminous, bejeweled girdle. The collar rises to the chin in the fashion of the time, and the sleeves are in contrast to the rest of the body by their width and floor-trailing length. Next to this figure is another suitor dressed in a long *houpplande,* while in the middle ground one sees a short version of the same garment, characterized by a rising collar, a cinched waist, and ample, falling sleeves. All these personages have protruding chests due to the wearing of a *pourpoint,* and the configuration of their garments is quite similar, in spite of their differences. Their upper torsos are magnificently broad, slim waists are emphasized, and the lower bodies are attenuated.

The garment's drawing closer to the body reduces the volume that previously gave the wearer an increased spatial importance and the ease of movement inherent in all ample garments. But, as the chroniclers noted, the body reduced to its center was endowed with a strange proliferation at its extremities that it re-deployed in the surrounding space. A manuscript of the works of Guillaume de Machault[25] shows how the body in the *pourpoint* somehow prolonged itself outward through the *poulaines,* the *chaperons* whose tips are sometimes tied into a knot, and the *coudières.* The slashing seen at the edges of the garments, strange, vegetal excrescences that become commonplace at the turn to the fifteenth century, play with the borderline that separates it from the exterior world.

Around 1400, the *chaperons* become taller, turning with increasing frequency into turbans whose fragile equilibrium bring a noteworthy elegance to the silhouette, to which a figure in a miniature of the Boccacio manuscript just cited bears witness. To an even greater degree, this also applies to the turbans represented in the New Year's banquet scene that opens the *Très Riches Heures* of the Duke of Berry.[26] The *carcaille* collars rise to the chin, drawing the garment generally upward as the novel length pulls it down. The sleeves, above all, became privileged supports for the ornamentation that was manifested especially in the superimposition of pieces, and consequently, in the play of the garments above and below. The body dressed was thus the locus of tension between movement and the different volumes that animated the silhouette in which the general, fifteenth-century style configuration had evolved, toward a masculine body elongated in its lower parts and with imposing shoulders. A miniature from a manuscript produced in Bruges between 1460–1470,[27] representing an

amorous couple, is an example of the formalism of the fashions of the second half of the fifteenth century. The male body, which had then lost its
padded chest but whose shoulders were considerably enlarged, was made
lighter in its lower parts through the constriction of the waist and the long
legs terminating in *poulaines,* forming a sort of triangle with its point
pointed downward. In the female dress, on the contrary, the volume is in-
verse. The bust was in effect very small, the high headdress elongated, and
the waist was high, while the lower body was lost in the considerable volumes of fabric. Face and bust were feminine, and shoulders were masculine zones: the sexual differences were inscribed on the garments that
outlined the positions of the body, invested by desire or power.

The moralists at the end of the Middle Ages could not stop themselves
from denouncing women's trickery and lasciviousness as invested in their appearance. Vying with this notion, however, the images show us that the vestimentary transformations of the time were the domain of men. The new
fashion, which took a military garment as its model, exalted the male body
while the female appeared less as a dangerous seductress than as an "always
nude" body, subject to masculine desires and conforming to its nature of
"being weak" by emulating again the traditional Christian morality. Illustrating this is a miniature displaying a couple that has tentatively been identified
as portraying Paris and Helen.[28] The man wears a short *houpplande* cinched
at the waist, its wide sleeves increasing the volume of his body and seemingly
enveloping his female companion with the impressive mass of fabric. The latter presents a slender silhouette in a low-necked gown devoid of any ornament. The materials and colors used accentuate this contrast. He is rendered
in a luminous and dense blue, and with a fur lining that confers a velvety
thickness to the garment. For her, there is a green gown mixed with yellow,
and with suddenly terminating edges lacking borders or lining as a finish.
While the fur discretely evokes the idea of the inherently savage part of the
male individual as well as his taming, the feminine garment exposes a perpetually open, vulnerable body, so to speak, less menacing than menaced. The
elegant and fur-clad gentleman, who asserts himself in a dominating manner
on the sleek and nude creature he holds in his arms, demonstrates how much
the modes of dress are part of the male power strategies. It is only in the
nineteenth century that men will relinquish to women the opportunity of
making a statement through their appearance, which, by this date, has in any
event ceased to be a reflection of being in command.

Notes

1. For a recent point of view, see Stella Mary Newton, *Fashion in the Age of
   the Black Prince* (Woodbridge: Boydell Press, 1981); and Odile Blanc, *Pa-*

rades et Parures. L'invention du corps de mode à la fin du Moyen Âge (Paris: Gallimard, 1997).

2. Anonimo Romano, *Cronica,* ed. G. Porta (Milan: Adelphi, 1981).

3. Jean de Venette is considered the last of the contributors to the chronicle in Latin of Guillaume de Nangis. See *Chronique de Guillaume de Nangis, avec les contunuations de 1300 a 1368,* ed. H. Géraud for Société de l'Histoire de France, 2 vols. (Paris, 1843).

4. See Gilles le Muisit, *Chroniques et Annales,* ed. H. Lemaître for Société de l'Histoire de France (Paris, 1906).

5. *Chronica Johannis de Reading et Anonymi Cantauriensis* (1346–1367), ed. J. Tait (Manchester, 1914).

6. See G. Villani, *Cronica,* ed. G. Dragomanni, 4 vols. (Florence, 1844–1845).

7. Tileman Ehlen von Wolfhagen, *Limburger Chronik,* ed. A. Wyss in the *Monumenta Germanie Historiae* (*Scriptores qui vernacula lingua usi sunt*) t.IV/ 1 (Hannover, 1883).

8. Paris BnF, ms Fr. 2813, *Les Grandes Chroniques de France,* Paris, ca. 1375–1379. See Anne D. Hedeman, *The Royal Image. Illustrations of the Grandes Chroniques de France, 1274–1422* (Berkeley-Los Angeles-Oxford: University of California Press, 1991).

9. Chantilly, Musée Condé, ms. 65, *Très Riches Heures du duc de Berry,* Paris, (1410–1416), f. 48.

10. See R. Barthes, *Le système de la mode* (Paris: Le Seuil, 1967).

11. M. Thomas, *L'Âge d'or de l'enluminure* (Paris: Vilo, 1983) (New York: George Brazillier, 1979). This work reproduces several of the images referred to here. See also Charles Sterling, *La peinture médiéval à Paris. 1300–1500* (Paris: Bibliothèque des arts, 1987); and the classic M. Meiss, *French Painting in the Time of Jean de Berry,* 3 vols. (London: Phaidon Press, 1967–1974).

12. J. Huizinga, *L'automne du Moyen Âge* (1919; Paris: Payot, 1980), 30.

13. See, for this turbulent category that plays a fundamental role in the adoption of the new fashions in the medieval society, the works of Georges Duby and in particular his "Les 'jeunes' dans la société aristocratique," *Annales* (ESC, 1964).

14. Rome, Biblioteca Apostolica Vaticana, Reg. Lat. 1505, Benoît de Sainte-Maure, *Roman de Troie,* f.50.

15. Reproduced in M. Laclotte and D. Thiébaut, *L'Ecole d'Avignon* (Paris: Flammarion, 1983), 29.

16. Paris, Bibliothèque Sainte-Geneviève, ms. 1029, Barthélémy l'Anglais, *De proprietatibus rebus,* Toulouse, ca. 1350, f. 8v.

17. New York, Metropolitan Museum of Art, Cloisters Collection, *Psalter of Bonne of Luxembourg,* Paris 1348–1349. Reproduced in Sterling.

18. The joint study of texts and images disposes one to think that the short garment, contrary to another, tenacious idea, cannot be reduced to a single "*pourpoint,*" but that there were several variations on the theme during 1340 and through the beginning of the fifteenth century.

19. Paris, BnF, ms. Fr. 1586, Guillaume de Machault, *Œuvres,* Paris, ca. 1350–1355, f.23, 52, and 55 reproduced in Sterling.

20. The *pourpoint* is a garment made of layered fabrics, with wadding of silk or cotton, held together by stitching. Hence its name from Latin *perpungere,* "piercing through stitching." It covers the top of the body under a suit of armor, also protecting it against injury from the metal.

21. " . . . et maintenant ils (les *pourpoints*) sont plus en cours que autres vestements, et par ce y convient plus de ouvriers et pourveoir aus choses selonc la mutation des temps" (Patent letters of Charles V, cited by R. de Lespinasse, "Les métiers et corporations de la ville de Paris" in *Histoire générale de Paris,* tome III, 184–5.

22. Benesch de Weitmühl, *Chronicon,* ed. Pelzel-Dobrowsky in *Scriptores rerum Bohemicarum,* tome II (1784), 22 ff.

23. La Haye, Museum Meermano-Westreenianum, MS 10 B 23, *Bible Historiale de Jean de Vaudetar,* Paris, 1372, f.2; reproduced in Sterling. Other examples may be seen in the works of Guillaume de Machault, Paris, BnF, ms. Fr. 1584, f. D.

24. Paris, BnF, ms. Fr. 12420, Boccace, *De cleres et nobles femmes,* Paris, around 1402, f. 12. Reproduced in Blanc. On this manuscript see also B. Buettner, *Boccacio's Des cleres et nobles femmes. Systems of Signification in an Illuminated Manuscript* (Seattle: University of Washington Press, 1996).

25. See note 19.

26. Chantilly, Musée Condé, ms. 65, f. 1v. Reproduced in Blanc.

27. Paris, Bibliothèque de l'Arsenal, MS 5072, Renaut de Montauban, around 1404–1409, f. 35. Reproduced in Blanc, fig 23.

28. Paris, BnF, ms. Fr. 606, Christine de Pisan, *L'Espistre Othea,* Paris, around 1404–1409, f. 35. Reproduced in Thomas.

CHAPTER 10

UNRAVELING THE MYSTERY OF
JAN VAN EYCK'S CLOTHS OF HONOR:
THE *GHENT ALTARPIECE*

Donna M. Cottrell

The sumptuous textiles depicted by the Flemish painter Jan van Eyck, active 1422–1441, include over 50 patterns, dozens of colors, and three types of fabric. Studies by this author, as well as others, have demonstrated that not only did the master distinguish between Flemish wools, Italian velvets, and *lampas* weave silks, but each was also characterized by the way in which it was employed. *Lampas* silks were used exclusively as cloths of honor, whereas wools and velvets were fashioned into garments and decorative items.[1] Further, specific adornments for the wools, and special categories of velvets informed Jan van Eyck's viewers of the portrayed figure's particular status in the hierarchy of the secular or heavenly court.[2]

Among Jan van Eyck's treasury of extraordinary textiles are seven cloths of honor, three of which are portrayed in the *Ghent Altarpiece* of 1432.[3] The altarpiece may be described as an encyclopedia of Christian history addressing the Incarnation, the sacrifice of Christ, and mankind's salvation. Commissioned by Joos Vijd and Elisabeth Borluut for their private chapel at the parish church of St. John the Baptist, the theme of the altarpiece was based on the liturgy for the Feast of All Saints and the Book of Revelation.[4]

In the upper register of the opened altarpiece, Jan van Eyck seated the Virgin, Christ, and John the Baptist before single panels of fabric known as testers or dorsers. Often referred to as "cloths of estate," the Ghent cloths of honor mirrored the arrangement and use of special textiles for royal seats of authority.[5] Still, it is of paramount importance to recognize that Jan van Eyck's cloths of honor were reserved for the enrichment of sacred thrones, and that they functioned both as sacred object and sacred text.

These concepts are not entirely new, yet the scope of Jan van Eyck's creativity in the Ghent cloths of honor, as well as their iconographic importance, have not been fully explored.

The Cloth of Honor and Its Significance

Rebecca Martin, in her study of medieval textiles, defined a cloth of honor as "a curtain of precious fabric suspended behind a saint as a sign of veneration," a motif that "reflect[ed] the way luxurious fabrics were used in courtly settings. . . ."[6] While this statement is true, cloths of honor functioned as more than a sign of veneration for the secular or religious person presented before it. Indeed, it is impossible to understand fully the significance of any cloth of honor without first giving careful consideration to the meanings associated with the sumptuous textiles used to create these majestic environments. These include the special "honor of cloth" in the secular realm, textiles' special connections to the Divine in the religious realm, and sacred ideas traditionally associated with imperial cloths of honor.

The "honor of cloth" refers specifically to the reverence bestowed upon luxurious textiles in the Middle Ages and the Renaissance. Initially, this honor seems to have been the result of the West's contact with Byzantium. There, imperial courts literally sparkled with textiles of silk and precious metals. Garments, thrones, palaces, and even horses displayed their splendor. Western ambassadors, kings, crusaders, and pilgrims often made their way home with a sampling of the precious materials.[7] However, by the twelfth and thirteenth centuries, luxury textiles became more widely available in the West because of growing economic wealth, Western control of coastal trading cities, and the ingenuity of Italian shipping magnates. The cherished materials from Byzantium, China, and the Middle East, and later from production sites in Sicily, Spain, and Italy, quickly came to play significant roles in the Western social, economic, and political arenas.[8]

Kings, dukes, and the wealthy elite, the only members of the secular realm who could afford such elegant textiles, wasted no time in employing them to their social and political advantage. They used the special fabrics to fashion garments, to adorn their courtly domains, and to advertise their special privilege at elaborately staged public events. They also presented lavish textiles as gifts to soon-to-be relatives, as pious donations to the church, as diplomatic gifts, and as leverage to gain political favor. Inventories and chronicles, especially those of the French kings and the Valois dukes, including Philip the Good, Jan van Eyck's employer, are filled with descriptions of textile finery exploited for these purposes.[9]

The aristocracy's use of luxury textiles for financial purposes also contributed to their high esteem. Because of their substantial value and gold and silver thread, they were collected in large quantities. Frequently bequeathed as inheritances, when times demanded, the textiles were also used as currency.[10] In fact, the close relationships between the courts and the textile merchant families were not based solely on the aristocracy's desire for luxury textiles, but upon the textile merchants' banking ties. Indeed, court documents betray the fact that the textile merchants' lending ability frequently had a direct impact on a court's ability to survive disasters, launch building campaigns, ransom relatives, wage war, and host elaborate festivities.[11]

While the honor of cloth in the secular realm stemmed from its many social, financial, and political roles, the religious realm added a new dimension to its reverence, a connection to the Divine. Priests and popes rivaling their Byzantine and secular counterparts for status and power, fashioned the sumptuous fabrics into ecclesiastic garments, and paraded their cache of special materials in the pageantry of feast day celebrations. Perhaps more importantly, they wrapped sacred relics in gold and brocaded cloth, and generously draped costly silks over precious vessels, altars, symbolic tombs, and behind venerated sculpture. The majestic textiles thus became inseparably associated with Christ, the saints, sacred events, and sacred places.[12]

These sacred connections were strengthened by the mindset of the Church. For example, the exotic textiles woven with gold, silver, and brilliantly-colored threads were believed to possess "special" properties of light—properties that not only reflected the essence or presence of Divine light, but could direct one's thoughts to the Divine in mystical meditation. This Neoplatonic view, that physical elements shared the essence or "light of God," was expressed by St. Augustine early in the Middle Ages, but similar ideas were cultivated in later centuries by Hugh of St. Victor, Rupert of Deutz, and Abbot Suger.[13] Additional connections to the Divine were encouraged by placing embroidered panels with the stories of Christ, Mary, and the saints onto the patterned silks, especially those silks that possessed a veritable menagerie of real and fantastic birds and beasts. All were popular subjects in the numerous bestiaries and exegetical texts that employed the images as Christian allegories.[14]

Textiles' connections to the Divine were also promoted by apocryphal and biblical accounts of special cloth. Celebrated stories included those of the veil of Saint Veronica, the "royal scarlet" woven by Mary, and of the miracle-working fragments of garments belonging to the saints. Sermons, parables, and metaphors extended the sacred aura of textiles. For example, Mary was said to have "clothed" Christ in her womb, and often Mary was

compared to a "curtain" that "hid and revealed" the Divine.[15] The cumu-
lative result of these religious associations was that the sumptuous fabrics
came to possess an innate aspect of sacredness about them. Therefore, the
awe-inspiring textiles provided a direct link to the heavenly domain.[16]

The sacred ideas traditionally associated with imperial cloths of honor
also provide important keys to understanding the Ghent presentations of
Mary, Christ, and John the Baptist. Throughout history high-ranking secular and religious leaders surrounded their thrones with rare and costly
textiles.[17] They were employed as throne adornments (throne drapery,
cushions, or other decorative items), curtains of honor, testers, and/or
canopies.[18] Yet, regardless of their form or arrangement, their display about
the throne always signaled divine kingship and supreme authority.

These connotations developed in several ways. First, the use of precious
textiles in royal presentations served "to elevate and to separate" the enthroned figure from the general populace. Second, their display asserted the
ruler's "equality and association with the divine" who was believed likewise enthroned in the heavens. The use of special fabrics, often glittering
with gold and silk threads, also reinforced the interpretation of the throne
environment as an "otherworldly" setting. Moreover, the use of cloths of
honor as backdrops, curtains, and canopies, implied the "divine conceal-
ment or revelation" of the figure seated within its boundaries, and therefore, marked the space as "sacred." In short, the use of special textiles to
enrich a throne environment affirmed the ruler's "divine right" to rule as
cosmocrator of an earthly realm.[19]

Throughout the Middle Ages and the Renaissance, the display of sumptuous cloths of honor continued to be viewed as symbols of royalty, divine
kingship, and of supreme authority. Their interpretation as such was applied to religious imagery as well, and, in each respect, these traditional interpretations may be applied to Jan van Eyck's cloths of honor in the *Ghent
Altarpiece*.[20] There, the three most holy and highest ranking figures of the
heavenly realm were enthroned above all humanity, blessed and crowned
by the Divine, surrounded and isolated by magnificent textiles, yet revealed
to all who contemplated the mysteries of what was represented.

The Ghent Cloths of Honor Patterns and Inscriptions

To describe the textiles, patterns, and inscriptions of Jan van Eyck's cloths
of honor is not an easy task. One must acknowledge that the painted depictions are at least one step away from reality. Further, there must be a
cautious reliance on past scholarship since many of the details were once
obscured from view by layers of old varnish and repairs. Additionally, terminology must be scrutinized, for often scholars have been more con-

cerned with the textiles' visual impact than their technical construction and patterns.[21] Today, scholars are challenged by paint losses and further restorative alterations.[22]

Perhaps it is most fitting to begin with Christ's cloth of honor. Few art or textile historians have failed to remark on its rich green fabric, gold-brocaded motifs of the pelican in piety, and banners inscribed with "IH-ESVS.XPS." Yet, the textile and patterns so carefully woven by Jan van Eyck have much more to reveal.

On careful inspection of Christ's cloth of honor, it becomes apparent that it is "made" from a single piece of fabric. And, judging from the breadth of Christ's shoulders, it would seem that it is a double-width textile, a special cloth indeed.[23] Also distinctive is that the cloth of honor seems to be mystically suspended behind the seated Christ, as there are no visible supports or tethers. In fact, the tester falls to the tiled floor unmarred by breaks or folds in the material.[24]

The details of the pattern and the surface texture of the textile, rendered palpable by Jan van Eyck, reveal that the fabric is an extraordinary *lampas* weave silk. Both the dark green ground and the background pattern of lighter green ivy are distinguished by diagonal hatching marks signaling the complexity of the weave.[25] A second set of diagonal hatching marks covers the textile's primary motifs of the pelican in piety and grape clusters, indicating they are added woven elements.

The primary motifs in Christ's cloth of honor coalesce to form rectangular pattern blocks of equal size.[26] These pattern blocks are, in turn, arranged in horizontal rows with a half-repeat—that is—the second row of the pattern is offset by one-half pattern from the row above it. Measurements indicate that six pattern blocks are equally spaced along the vertical plane. However, the spacing of the horizontal pattern is irregular. If the equally-spaced blocks to the left and to the right of Christ were to be continued across the width of the textile, two blocks would not fill the space behind Christ, and three blocks would be too large for the available space.[27] Jan van Eyck, in this instance, seems to have stretched the horizontal pattern in order to permit a fuller view of the details, and to balance the compositional elements within the textile and the panel.

The primary motif in Christ's pattern blocks is that of the adult pelican with its three young nestled in a boat-shaped nest of slender leaves, stylized ivy, and delicate flowers. The adult pelican bends her head forward pricking her breast, while her young strain to catch the blood flowing from her wound. A flowering tree grows upward from the pelican's nest and divides into three branches. The branches to the left and to the right each support a cluster of grapes and a single grape leaf, while the smaller center branch sprouts a variety of blossoms: an upside down heart-shaped bud, a flower with four petals, and a stylized fleur-de-lis blossom with a heart-shaped base. Similar blossoms are scattered about the pattern block to "fill-in" the design.

Completing the pattern, and forming an arch directly above the family of pelicans, is a banner inscribed in red lettering. The inscription, "IH-ESVS.XPS" or "Jesus Christ," is written in the Roman and Greek alphabets. Complete inscriptions can be seen clearly in the pattern block immediately to the right of Christ's blessing hand and on the two banners located at the far right margin of the textile. Only glimpses of text may be seen in the remaining banners.

The Virgin's cloth of honor shares several of the features of Christ's textile. For instance, it is fashioned from a single piece of fabric,[28] reveals no evidence of its means of suspension, or any visible folds or breaks in the material. Finally, its primary pattern blocks also display diagonal hatching.

Nevertheless, Mary's cloth of honor possesses several distinguishing characteristics. (Figure 10.1.) First, although her cloth of honor seems to have been made from a *lampas* weave silk because of the complexity of its design and pattern, the pure white ground does not display the diagonal hatching found over Christ's ground weave. Nor are the silver floral motifs scattered between Mary's pattern blocks marked by diagonal hatching.[29] Perhaps Jan van Eyck intended this "absence" to signal a different ground weave (e.g., plain or satin v. twill), and a different "brocading" technique (e.g., continuous v. discontinuous). In any event, the absence of the hatching pattern creates visually a sense that Mary's cloth of honor is a lighter-weight and softer material than Christ's.

The structural arrangement of the pattern within the Virgin's textile differs as well. Mary's pattern blocks are placed in a strict grid format with the pattern aligned horizontally *and* vertically. Further, measurements of the pattern blocks indicate that they are slightly larger and squarer than those in Christ's cloth of honor. Also, while six equally-sized blocks fit vertically as in Christ's textile, five pattern blocks fit comfortably across the width of the Virgin's tester. No adjustment of the horizontal pattern seems to have been necessary.

The primary motifs within Mary's pattern blocks consist of cottonball-like clouds, radiant sunbeams (or rain), a forested haven sheltering a unicorn, and a banner inscribed with red letters. Careful study of the recumbent unicorn reveals that his legs enfold a small tree. He also sports a spectacular twisted horn that extends the entire length of his torso.

An inscribed banner, positioned almost as a cradle for the unicorn, completes the primary pattern. Of the four banners on Mary's tester (three to her left and one to her immediate right) only two possess complete inscriptions. In the past, scholars have characterized the lettering as Greek,

Figure 10.1   The Virgin Mary. Detail, cloth of honor. The *Ghent Altarpiece*, Jan van Eyck, 1432. (Copyright IRPA-KIK, Brussels. Used by permission.)

Figure 10.2   Illustration by Friedrich Fischbach of M. Coxcie's copy of the pattern of the Virgin Mary's cloth of honor. (By permission of Dover Publications.)

Arabic, Hebrew, Saracenic, legible but inexplicable, and illegible.[30] Indeed, multiple restorations of this textile have rendered many of the letters fragmentary at best.[31] Yet, based on the clearly legible inscriptions in Christ's cloth of honor, and elsewhere throughout the *Ghent Altarpiece,* it seems unreasonable to assume that Jan van Eyck "scribbled" unintelligible letters on the Virgin's banderoles. This theory is supported by the fact that the letters share a general form, and that the order of the letters appears to be the same in each of the four banners. Additionally, the last two letters of the inscriptions can be reconstructed with little difficulty. They are "XP" (Chi Rho), a variation of Christ's monogram.

The first "word" of the banners, however, poses a significant challenge. It does not resemble the "IHESVS" of the Christ banderole.[32] Moreover, it is difficult in many instances to determine how many letters are in the first word, or whether the letters are written in the Greek or Roman alphabet. Adding further to the puzzle is the suspicion that Jan van Eyck may have used an abbreviated phrase. Thus, the first "word" in the Virgin's banner could be more than one word, in more than one alphabet, or in more than one language.[33]

This riddle is partly solved by Fischbach's reproduction of the Virgin's textile pattern in his *Geschichte der Textilkunst*.[34] (Figure 10.2.) His reproduction, however, is based on Michiel Coxcie's copy of the *Ghent Altarpiece* made for Philip II in 1557, not Jan van Eyck's original panel. Nevertheless, Fischbach's illustration reveals that while Coxcie dramatically simplified Jan van Eyck's painted brocade, Coxcie retained his primary motifs. It is suggested Coxcie also *transcribed* from Jan van Eyck's banners the phrase "KREUZ.XPS" or "Cross of Christ."[35]

Would Jan van Eyck have written these words on the Virgin's banners in the *Ghent Altarpiece?* The phrase is applicable to the Virgin (see discussion below), and the lettering is close to the fragmented remains of the inscription. However, "KREUZ.XPS" does not seem to be a perfect match. As noted above, Christ's monogram in Mary's banner is "XP" not "XPS." Additionally, the first letters of Mary's banners more closely resemble "XP" (as in Christ's banner), not "KR." Finally, the fourth letter of the first word does not appear to be a "U." Therefore, only Coxcie's "E," "Z," and final "XP" seem to fit without effort.[36]

Despite this difficulty, Coxcie's transcription may still retain its validity. What if Coxcie *translated* the first word of Jan van Eyck's inscription into its Flemish counterpart? Moreover, what if Jan van Eyck spelled *kreuz* phonetically using the Roman and Greek alphabets?[37] The first word of Mary's banner would then become "Chi Rho E Omega Z," or perhaps "Chi Rho Epsilon Omega Zeta" (XPEΩZ).[38] Phonetically, this spelling would produce the appropriate Flemish vocalizations. Certainly, there is nothing unusual about this type of word construction for Jan van Eyck. The use of a mix of alphabets and languages is entirely consistent with the construction of other words and phrases in the *Ghent Altarpiece*. It is also clearly within the artist's spelling habit as evidenced by his own motto "ALS IXH XAN," and inscriptions in his portraits of Tymotheos, Jan de

Figure 10.3 St. John the Baptist. Detail, cloth of honor, The *Ghent Altarpiece*, Jan van Eyck, 1432. (Copyright IRPA-KIK, Brussels. Used by permission.)

Leeuw, and Margaret van Eyck.[39] Indeed, the use of the Greek alphabet could explain the unusual appearance of the third, fourth, and/or fifth letters in the Virgin's banners. It would also explain Coxcie's ability to "reproduce" Jan van Eyck's inscriptions, as well as our present-day difficulty in reconstructing them.

John the Baptist's cloth of honor has been the least studied of the Ghent textiles. (Figure 10.3) Perhaps this has been due to the serious damage the panel has suffered during its lifetime. Restorative efforts and paint losses are apparent even at moderately close range. Pattern details and colors are blurred across the entire fabric, and especially along the lateral margins of the panel.[40]

Historically, comments have been restricted to the fact that John's cloth of honor mimics Mary's except for its red ground and bluish-green floral sprays. In fact, a comparison of the two textiles reveals it is constructed of a similar *lampas* silk. For example, John's ground weave and ancillary floral design are not distinguished by diagonal hatching. Also, John's cloth of honor is mystically suspended and displays no folds or breaks in the material. Additionally, the pattern blocks of John's tester are the same size as Mary's and are arranged in the same grid format. Lastly, John's primary motifs repeat Mary's.

Although the color of the material and floral sprays distinguish John's tester from Mary's, two additional differences should be noted. First, John's flowers have distinctly heart-shaped petals. Second, John's banners are fewer, and his inscriptions seem to be more cryptic than Mary's.

Unfortunately, today the lettering in John's banderoles is quite confused. Adding to the mystery of their inscription, each is cropped—either by John's robe or by his hand. Still, several clues to their message remain. Careful examination of the banderoles reveals that the "first part" of the inscription is provided in the two banners located by John's raised right hand. The "last part" of the inscription is presented in the banderole near John's left shoulder. Therefore, Jan van Eyck provided contemporary viewers (and us) with a complete inscription, if only the fragmented letters can be pieced together.

Analysis of the first two letters confirms that, although the individual strokes for the letters are now widely separated, they may have formed "XP" (Chi Rho) as in the Virgin's banners. The third letter in John's banderole, an "E," is clear presently only in the lower right-hand banner. However, the last three letters may be reconstructed more easily. They are "ZXP." If correct, the "E" and "ZXP" would also duplicate the lettering sequence in the Virgin's banners.[41] Based on these observations, it seems reasonable to suggest that John's inscription repeated the message of Mary's banderoles—that is "XPEΩZ.XP" or "Cross of Christ."

The Sacred Text of the Ghent Cloths of Honor

Despite the tomes published on the *Ghent Altarpiece,* little attention has been focused on the specific iconography of its cloths of honor.[42] Generally, comments have been restricted to their symbolic colors, and to the motifs of the pelican, grapes, and unicorn as they relate to Christ, and/or to Mary. While the green, red, and white colors of their testers were most surely interpreted as symbolic of eternal life, martyrdom, purity, and the Eucharist,[43] it is the figurative and literal text of the Ghent cloths that demands our attention here.

The combination of image and actual text in the Ghent textiles is unique to Jan van Eyck's cloths of honor, although his habit of encouraging interaction between imagery, text, and textual sources is not.[44] Looking specifically at Christ's cloth of honor, this interaction is clear. Prominently displayed within the main pattern, the pelican stands in piety with its young. Since the second century, this bird had a special connection to Christ's Passion. Ecclesiastic and secular texts repeatedly drew parallels between the adult pelican's sacrifice for its young and Christ's sacrifice for the salvation of mankind.[45] Artists portrayed the allegorical image on altars, liturgical vessels, in church architecture, and over Crucifixes throughout the Middle Ages and the Renaissance. Similar sources supplied references for the Eucharistic connotations of the grape clusters within the primary pattern of Christ's cloth.[46]

Yet, not to be ignored, are the less prominent motifs within Christ's tester. They too seem to relate to the theme of Christ's sacrifice and mankind's salvation. Of primary importance is the single tree that grows from the pelican's nest of ivy. In medieval devotional gardens, the single tree represented the Tree of Life in the Garden of Paradise, the very tree from which the cross of Christ was made.[47] The application of this interpretation to the tree in Christ's textile seems to be entirely appropriate, especially if it is recognized that the tree is rooted in the pelican's nest of ivy (a symbol of the Incarnation),[48] that it blossoms forth clusters of grapes, and that it splits into three branches (the shape of a cross). Christ's banner inscription above these images, therefore, serves not only as a nameplate, but underscores the sacrificial and Eucharistic relationship of the imagery to Christ.

The figurative text of the Virgin's cloth of honor seems to convey similar messages. For instance, the primary motif of the Virgin's textile, the unicorn, is the Christological symbol of the Incarnation of Christ according to medieval exegesis and bestiaries. As such, the unicorn represents Mary's role in the salvation of mankind as the bearer of the Logos.[49] However, these sources also associate the unicorn with the Passion of Christ, for the spiritual beast was hunted and killed for its redemptive power, as was Christ. Moreover, allegorical references to Christ propagated by the Church Fathers and subsequent writers, promoted the idea that the single horn of the unicorn symbolized the oneness of the Father and the Son, the unity of faith, and the single power shared by the Father and the Son. The single horn of the unicorn was also equated to the cross of Christ.[50]

The smaller details within Mary's textile also may have served a dual purpose. For example, the forested haven of the unicorn was often compared to the *hortus conclusus* of the *Song of Songs*. Consequently, its enclosure became synonymous with Mary's virginity and the Incarnation. Indeed, it is possible that the clouds and sunbeams (or rain) above the forested haven referenced the Incarnation as well. Yiro Hirn, in particular, cites a metaphor that hails Mary as "clouds enclosing the Sun (Son)" bringing forth "rain (or blessings) to earth."[51] Conversely, the *hortus conclusus* was the place of the unicorn's sacrifice and, therefore, most likely provided an equally powerful reminder of Christ's sacrifice.

The single tree growing within the unicorn's enclosure, and seemingly grasped by its folded legs, deserves special mention. Like the single tree growing from the pelican's nest in Christ's cloth of honor, the solitary tree within the unicorn's forested haven surely referenced the Tree of Life and the cross of Christ. Two details within Mary's textile seem to emphasize this connection. The first is Mary's banner proclaiming Christ's sacrifice on the cross. The second is that the tree growing within the enclosure is a pomegranate tree. Its fruit not only symbolized Christ and immortality, but its blood-red seeds and juice were often likened to the blood of Christ.[52]

One final botanical detail within Mary's cloth of honor remains—the silver flowers surrounding the primary pattern blocks of the textile. Although only several petals are visible, their positioning seems to indicate that each flower would have had four petals. The ogival-shaped petals most closely resemble those of the stock gillyflower, a flower sometimes associated with the Virgin, but more frequently associated with Christ. Its nickname, "nail flower," derived from its four petals that reminded one of Christ's cross.[53] In devotional images, it was commonly represented around or within the mystical unicorn's enclosure.

Although John's role within the altarpiece has been studied extensively, scholars have not commented on the specific relationship of the textile pattern to John.[54] Perhaps this has been due to the perception that the presence of the unicorn is "unusual" in his textile, or that the "meaning" of John's imagery is no different than that in Mary's cloth of honor. Yet, textual evidence confirms that the unicorn's appearance in John's tester is not unusual, nor do the motifs necessarily convey the same message as those in Mary's.

Foremost, the appearance of the unicorn in John's tester may be explained by John's position and function within the altarpiece. Traditionally the attribute of John is the lamb. However, in this instance, the Lamb of God (Christ) is enthroned to John's immediate right. Further, John's gesture recalls his proclamation, "Behold the Lamb of God," a message also repeated on the altar antependium in the lower center panel of the altarpiece. Additionally, the symbol the "Lamb of God" appears on that altar. Considering these factors, Jan van Eyck may have perceived the use of a lamb in the textile pattern as redundant or confusing.

Still, there may have been other factors that influenced Jan van Eyck's choice of the unicorn for John's cloth of honor. Although the unicorn and the lamb were employed as interchangeable symbols for Christ,[55] the symbols had distinct connotations—the unicorn symbolized the Incarnation *and* the sacrifice of Christ, while the lamb was interpreted almost exclusively as a symbol of Christ's sacrifice. Thus, Jan van Eyck's use of the unicorn in John's textile may have been intended to encourage contemporary viewers to contemplate both events rather than Christ's sacrifice alone. Moreover, this substitution enabled John to point to the means of redemption (Christ), while at the same time remind the viewer symbolically of the sacrifice that made that redemption possible. The connection seems to be emphasized by John's banderoles inscribed with "Cross of Christ."

Perhaps an equally important factor in Jan van Eyck's choice of imagery was the unicorn's ability to stress the relationship *between* Mary and John. For instance, both shared special spiritual births—Mary at the Incarnation of Christ, and John at the moment of Mary's greeting to his mother Elizabeth. Both also were special participants in the redemption of mankind—Mary as bearer of the Logos and John as witness to the Logos. Indeed, Mary and John function as deacon and subdeacon in the Eternal Mass by the very nature of their position beside Christ in the altarpiece. Moreover, Mary and John serve as the Bride and Friend of the Groom in the Mystical Wedding of the *Song of Songs,* an image alluded to by the forested haven of the unicorn, as well as by other details within the altarpiece.[56]

The identity of the flowers located between the pattern blocks in John's cloth of honor remains elusive. Only two petals of two blossoms are visible, yet their position seems to indicate that they would have four petals per blossom. Their heart-shaped petals most closely resemble the blossoms of the primrose family, a flower that was an attribute of the Virgin Mary, and a symbol of the Incarnation. As a "flower of the fields" in the *Song of Songs,* artists often included it within the enclosed garden of the mystical unicorn.[57] Its association with John the Baptist, however, is unclear, although its cruciform portrayal must have fostered contemplation of Christ's sacrifice.

The Reality of the Ghent Cloths of Honor

There is little doubt among textile experts that the textiles in Jan van Eyck's paintings reflect the silks produced in Italy in the late fourteenth and early fifteenth centuries. Italian production centers, were, in fact, responsible for nearly all of the silk fabrics sold in Europe, and especially those tendered in the largest silk market of the time, that of Bruges.[58] Yet, are Jan van Eyck's Ghent cloths of honor accurate reproductions of contemporary Italian silks, or do their designs and patterns reflect the artist's creativity?[59]

Traditionally, textile historians establish that a painted textile accurately reflects a contemporary fabric by matching the painted fabric with extant fragments or a "family" of extant patterns. While precise, the scarcity of fourteenth and fifteenth century textiles complicates this process. Nevertheless, the Ghent cloths of honor have some interesting relatives among those surviving fragments.

Two extant Italian silks, in fact, have long been associated with Christ's cloth of honor. Both textiles, cited by Marien-Dugardin in 1947, date to the late fourteenth century. One silk fragment displays horizontal bands of affronted pelicans in piety, spotted panthers, a variety of small flowers, grape leaves, and blossoming trees potted in vases. The second fragment, from a dalmatic, portrays the pelican in piety standing over a nest of three eggs. Each is isolated by ogival-shaped "walls" of grape clusters and grape leaves.[60] Still, while textiles such as these may have served as a model for the primary motifs and half-repeat layout of Christ's cloth of honor, even Marien-Dugardin acknowledged that the inscribed banners, ivy nests, and block format in Jan van Eyck's tester were not common features of the grape-leaf-and-cluster family and, therefore, must have been extracted from other contemporary silks. Indeed, from this author's study of extant and painted textiles, it appears that Jan van Eyck selected specific elements from a number of contemporary silks in order to create Christ's cloth.[61]

Interestingly, in contrast to Christ's tester, the cloths of honor behind the Virgin Mary and John the Baptist seem to have a firmer connection to contemporary *lampas* silks. As noted previously, their cloths of honor share a basic pattern and pattern arrangement. Not only does this repetition suggest that their cloths may have been modeled on an actual textile, but corroborating evidence seems to be provided by a family of extant fragments animated by similar creatures in similar naturalistic environments.[62]

One fifteenth-century Italian textile, of which several fragments survive, is remarkably close to the pattern of Mary and John's cloths of honor. It displays magnificent unicorns with long twisted horns resting beneath pomegranate trees. A stylized sun showers rays of light over the unicorn's paradise, while a small fence encloses its forested haven.[63] However, it is not an exact match. Comparisons reveal Jan van Eyck's painted pattern is arranged in a grid format rather than in a half-repeat. Further, extraneous secondary motifs, such as the rambling foliage, have been eliminated from the cloths of honor. Additionally, Jan van Eyck has enriched the habitat of his unicorns with a leafy groundcover, and naturalistic sunbeams and clouds. He has also replaced the fence with an inscribed banner.[64]

Perhaps what has been overlooked when considering whether the Ghent cloths of honor are accurate reproductions of contemporary patterns is their inter-relationship. Based on available textile evidence, it seems possible that Jan van Eyck modeled the three cloths of honor on a "unicorn" textile such as the one cited above for Mary and John. The unicorn textile provides a close match in imagery, and a straightforward layout of design elements. Additionally, because the most natural association of the unicorn is with Mary, perhaps Mary's cloth of honor served as Jan van Eyck's inspiration. It would have required only slight modification for the Virgin's cloth of honor, and only a few additional minor changes for use as John's tester (different symbolic colors, flowers, and a slightly altered banner). Jan van Eyck, then, could have created Christ's cloth of honor by appropriating pertinent imagery from other available contemporary textiles. Adjustments in Christ's pattern (half-repeats and slightly different-sized pattern blocks) would have insured the three cloths of honor complemented one another aesthetically as well as iconographically.

While the question of the reality of Jan van Eyck's *lampas* silks may remain difficult to answer, it is evident that the Ghent cloths of honor do not *faithfully* copy any known textiles. Moreover, it is evident that Jan van Eyck carefully selected the motifs for each of the testers to insure that they related specifically to the sacred images within the composition. For contemporary viewers, the textiles may have established a worldly point of reference, yet their purpose was always to direct them to "otherworldly" Christian beliefs.[65] To this author, the Ghent textiles were unquestionably created—both as a part of a coordinated symbolic program, and as a part of what Ward described as Jan van Eyck's "deliberate strategy to create an experience of spiritual revelation."[66]

In summary, the intricate patterns of Jan van Eyck's Ghent cloths of honor speak sermons. Their literal and figurative texts relate directly to the figure enthroned before them, and repeatedly stress the sacrifice of Christ and the promise of salvation. Their imagery derives from popular stories, devotional literature, sermons, parables, and metaphors that would have been easily recalled by Jan van Eyck's audience. How sophisticated those recollections may have been would have been dependent upon the enlightenment of the patron or the pious viewer. Still, regardless of the viewers' education, the importance of the precious textile setting, and the special qualities of the Ghent cloths of honor would not have been lost on his contemporaries—either as sacred object or as sacred text. And, understanding the meanings associated with the patterns, inscriptions, and textiles is paramount if we are to comprehend why Jan van Eyck took such care in the representation of his cloths of honor.

Notes

1. Donna M. Cottrell, "Birds, Beasts, and Blossoms: Form and Meaning in Jan van Eyck's Cloths of Honor" (Ph.D. Dissertation, Case Western Reserve University), 1998, 82–96, and "Jan van Eyck's Closet Iconography," paper presented in "*Medieval Textiles: Object, Text and Image,*" The 33rd International Congress on Medieval Studies, Kalamazoo, MI, May 1998. Lisa Monnas, "Silk Textiles in the Paintings of Jan van Eyck," in *Investigating Jan van Eyck,* ed. Susan Foister, Sue Jones, and Delphine Cool (Turnhout: Brepols, 2000), 147–62.

2. Cottrell, "Closet Iconography"; Robert Baldwin, "Textile Aesthetics in Early Netherlandish Painting," in *Textiles of the Low Countries in European Economic History,* ed. Erik Aerts and John H. Munro (Leuven: Leuven University Press, 1990), 32–40.

3. Cottrell, "Birds, Beasts," 64–75, 105–45.

4. Otto Pächt, *Van Eyck and the Founders of Early Netherlandish Painting,* trans. David Britt (London: Harvey Miller Publisher, 1994), 127; Carol Purtle, *The Marian Paintings of Jan van Eyck* (New Jersey: Princeton University Press, 1982), 16–21.

5. Penelope Eames, "Furniture in England, France, and the Netherlands from the Twelfth to the Fifteenth Century," in *Furniture History* 13 (1977): xvii, 1–276, esp. 74; Monnas, 152; Jeffrey Chipps Smith, "The artistic patronage of Philip the Good, Duke of Burgundy (1419–1467)" (Ph.D. Dissertation, Columbia University, 1979), 189.

6. Rebecca Martin, *Textiles in Daily Life in the Middle Ages* (Bloomington: Indiana University Press, 1985), 33.

7. John Beckwith, *The Art of Constantinople: An Introduction to Byzantine Art 330–1453* (New York: Phaidon, 1961), 93–94, 100–104; Christine V. Bornstein and Priscilla P. Soucek, *The Meeting of Two Worlds: the Crusades and the Mediterranean Context* (Ann Arbor: University of Michigan Museum of Art, 1981), 9, 17; J. P. P. Higgins, *Cloth of Gold: A History of Metallised Textiles* (London: Lurex Co., Ltd., 1993), 12.

8. Janet Ellen Snyder, "Clothing as Communication: A study of clothing and textiles in Northern French Early Gothic Sculpture (Portals)" (Ph.D. Dissertation, Columbia University, 1996), 503–14; John Kent Tilton, *Textiles of the Italian Renaissance: their history and development* (New York: Scalamandré Silks, Inc., 1950), 1–18.

9. Baldwin, 32; Higgins, 24, 33–4, 37; Florence Edler de Roover, "The Silk Trade of Lucca," *Bulletin of the Needle and Bobbin Club* 38 (1954): 28–48; H. Wescher, "Fabrics and Colours in the Ceremonial of the Court of Burgundy," 1850–56, and "Fashion and Elegance at the Court of Burgundy," 1841–48, *Ciba Review* 51 (1946); Michèle Beaulieu and Jeanne Baylé, *Le Costume en Bourgogne de Philippe le Hardi à la mort de Charles le Téméraire (1364–1477)* (Paris: Presses Universitaires de France, 1956), 27–30.

10. Beaulieu, 27; Bornstein, 15; Higgins, 24.

11. Leon Mirot, "Études lucquoise: Galvano Trenta et les joyaux de la couronne," *Bibliothéque de l'École des Chartres* 101 (1940): 116–56; Florence M. Edler, "The Silk Trade of Lucca during the Thirteenth and Fourteenth Centuries," Ph.D. Dissertation, University of Chicago, 1930, 96–9, 133–45; Raymond de Roover, *Money, Banking and Credit in Medieval Bruges: Italian Merchant-Bankers, Lombards, and Money-Changers* (Cambridge: Medieval Academy of America, 1948), 21–2, 39; V. Vermeersch, *Bruges and Europe* (Antwerp: Mercator, 1992), 188–93.

12. Y. Hirn, *The Sacred Shrine: A Study of the Poetry and Art of the Catholic Church* (1912; reprint Boston: Macmillan & Co., 1957), 145–74; Christa Mayer-Thurman, *Raiment for the Lord's Service, A Thousand Years of Western Vestments* (Chicago: The Art Institute of Chicago, 1975), 43–4.

13. Dieter Jansen, "Similitudo: Untersuchungen zu den Bildnissen Jan van Eycks" (Ph.D. Dissertation, Cologne University, 1988), 44–6; Millard Meiss, "Light as Form and Symbol in Some Fifteenth-Century Paintings," in *The Painter's Choice: Problems in the Interpretation of Renaissance Art* (New York: Harper & Row, 1976), 3–18; Erwin Panofsky, *"De Administratione" in Abbot Suger on the Abbey Church of St. Denis and Its Art Treasures,* 2nd ed. (New Jersey: Princeton University Press, 1979), 63–5.

14. George C. Druce, "The Mediaeval Bestiaries, and their Influence on Ecclesiastical Decorative Art," *British Archaeological Association Journal* New Series 25, pt. 1 (1919): 41–82, and 26, pt. 2 (1920): 35–79; Florence McCulloch, *Medieval Latin and French Bestiaries* (Chapel Hill: University of North Carolina Press, 1960), preface 7ff.

15. Ewa Kuryluk, "Metaphysics of cloth: Leonardo's draperies at the Louvre," *Arts Magazine* 64 (1990): 80–2, and *Veronica and her cloth: history, symbolism, and structure of a "true" image* (Cambridge: B. Blackwell, 1991), 71, 180–6; Hirn, 33–4, 163, 321, 454–5.

16. Martin, 12; Mayer-Thurman, 14.

17. Johann Konrad Eberlein, "The Curtain in Raphael's Sistine Madonna," *Art Bulletin* 65 (1983): 61–77; H. P. L'Orange, *Studies on the Iconography of Cosmic Kingship in the Ancient World* (New York: Caratzas Brothers, 1982), 135–6; E. Baldwin Smith, *Architectural Symbolism of Imperial Rome and the Middle Ages* (New Jersey: Princeton University Press, 1956), 4, 107–18, 151–5, 166–8, 197–8.

18. Cottrell, "Birds, Beasts," 27–30.

19. Per Beskow, *Rex Gloriae. The Kingship of Christ in the Early Church* (Stockholm: Almquist & Wiksells, 1962), 12–14; E. H. Kantorowicz, "Laudes re-

giae," *University of California Publications in History, Berkeley* 33 (1946): 56–9, 225–7, and *The King's Two Bodies: A Study in Medieval Political Theology* (New Jersey: Princeton University Press, 1957), 65, 93.

20. Moshe Barasch, *Imago Hominis: Studies in the language of Art* (New York: New York University Press, 1991), 25–6.

21. Brigitte Klesse, "Darstellung von Seidenstoffen in der Altkölner Malerei," in *Mouseion: Studien aus Kunst und Geschichte für Otto H. Foerster,* ed. Heinz Ladendorf (Cologne: M. duMont Schauberg, 1960), 218; Barbara Markowsky, *Europäische Seidengewebe des 13.–18. Jahrhunderts* (Cologne: Kunstgewerbe-Museum, 1976), 99–113; Donald and Monique King, "Silk Weaves of Lucca in 1376," in *Opera Textilia variorum Temporum: The Museum of National Antiquities Stockholm Studies* 8 (1988): 67–76.

22. Elisabeth Dhanens, *Van Eyck: The Ghent Altarpiece* (New York: Viking Press, 1973), 130–7; P. Coremans, *Les Primitifs Flamands: III. Contributions à L'etude des L'agneau mystique au laboratoire: examen et traitement* (Antwerp: De Sikkel, 1953), 100–1; J. R. J. van Asperen de Boer, "A scientific re-examination of the Ghent Altarpiece," *Oud Holland* 113 (1979): 143–5.

23. There are no visible "seams." Even assuming the cloth of honor is the same "width" as the panel (80 cm), the textile would be larger than the standard loom weave of one *braccia* or 59 cm; Lisa Monnas, "Contemplate What Has Been Done: Silk Fabrics in Paintings by Jan van Eyck," *HALI* 60 (1991): 103–13, and "Opus Anglicanum and Velvet: The Whalley Abbey Vestments," *Textile History* 25 (1994): 3–27; King, 67–8.

24. Monnas, "Silk Textiles," 152. Although referred to as draping high-backed thrones, the absence of breaks in the material confirms the testers are "suspended" behind the throne. The throne is most likely an X chair, one without a back and hidden by their generous garments; Eames, 181, 191.

25. Monnas, "Silk Textiles," 150; Brigitte Tietzel "Sein und Schein in Jan van Eyck's gemalten Stoffen," in *Festschrift für Brigitte Klesse,* ed. Ingrid Guntermann and Brigitte Tietzel (Berlin: P. Hanstein Verlag, 1994), 217–31; Lucy Trench, "Italian silks in fifteenth century Netherlandish painting," in *New Perspectives. Studies in art history in honour of Anne Crookshank,* ed. Jane Felon (Dublin: Irish Academic, 1987), 59–73; Anne E. Wardwell, "Italian Gothic silks in the museum collection," *Bulletin, LA County Museum of Art* 24 (1978): 6–23.

26. Measurements are based on author's observations and photographic details.

27. Monnas, "Silk Textiles," 148, agrees but does not comment specifically on this irregularity.

28. It is difficult to determine if the textile is "double-width." This panel measures 72.3 cm without the frame. The dispersion of the large pattern seems to indicate it could be two *braccia* in width.

29. To my knowledge this has not been noted previously.

30. Chanoine van den Gheyn, *L'interprétation du Retable de Saint-Bavon á Gand: l'Agneau Mystique des frères Van Eyck* (Ghent: n.p., 1920), 113; A. M. Marien-Dugardin, "Les draps d'honneur du Retable de l'Agneau Mystique," *Bulletin de la Societe Royale d'archeologie de Bruxelles,* 1947–48, 18–21; Ferdinand

de Mély, "Le retable de l'agneau des van Eyck et les pierres gravées talismaniques," *Revue archéologique* 14 (1921): 33–48; Hippolyte Fierens-Gevaert, *La Renaissance Septentrionale et Les Premiers Maitres des Flandres* (Brussels, G. van Oest & Cie, 1905), 176–220; Tietzel, 229; Monnas, "Silk Textiles," 152–3.

31. Coremans, 35–6, 45–6, 48–56, 64, 100–1; Van Asperen de Boer, 163–5; Elisabeth Dhanens, "Bijdrage tot de studie de de repentirs en oude overschilderingen op het Lam-Godsretabel van Hubert en Jan van Eyck," *Bulletin de l'Institut royal du patrimoine artistique* 15 (1975): 110–8.

32. Monnas, "Silk Textiles," 153.

33. Dana Ruth Goodgal, "The Iconography of the *Ghent Altarpiece,*" Ph.D. diss., 1981, University of Pennsylvania, 174, 325–7; Mély, 33–48; Jansen, "Similitudo," 10.

34. Friedrich Fischbach, *Die Geschichte der Textilkunst* (Frankfurt-am Main, n.p., 1883), pl. 75.

35. Ludwig Kammerer, *Hubert und Jan van Eyck* (Bielefeld: Verlag von Delhagen & Klafing, 1898), 11–13; J. Duverger, "Kopieën van het 'Lam Gods' Retabel van Hubrecht en Jan van Eyck," *Bulletin Koninklijke Musea voor Schone Kunsten,* Brussels 3 (1954): 51–68; Marien-Dugardin, 19; Monnas, "Silk Textiles," 152.

36. Author's observations.

37. I would like to thank Dr. Charles E. Scillia for his suggestion that the banner inscriptions may be Flemish spelled phonetically using the Greek alphabet and his invaluable assistance with the intricacies of the languages.

38. Due to editorial constraints, the Greek alphabet could not be printed. I believe the fourth letter is the capital Omega—the upside down "U" rather than the "W."

39. Charlene S. Engel, "Sator ara te: the Ghent Altarpiece cryptogram," *Revue Belge d'Archéologie et d'Histoire de l'Art* 62 (1993): 47–65; Dieter Jansen, "Jan van Eyck's Selbstbildnis—der Mann mit dem rotten Turban und der sogenannte Tymotheos der Londoner National Gallery," *Pantheon* 47 (1989): 36–48; Mély, 33–48; Paul Philippot, "Texte et image dans la peinture des Pays-Bas au XV et XVI siècles," *Bulletin Museés Royaux des Beaux-Arts de Belgique* 34–7 (1985–88): 75–86; D. de Vos, "Further notes on Als Ich Can," *Oud Holland* 97 (1983): 1–4.

40. John's panel measures 72 x 162.2 cm without the frame. Coremans, 45–8, 50–6, 63–4, 66–7; Van Asperen de Boer, 169; Roger H. Marijnissen, "Twee specificke paneelproblemen: de Johannes de Doper van het Lam Gods en Ruben's Kruisoprichting," *Bulletin de l'Institut royal du patrimoine artistique* 19 (1982–83): 120–32.

41. The fourth letter appears in all three banners, but can only be deduced by playing the word game "Hangman"; Monnas, "Silk Textiles," 153.

42. The exception is Purtle, whose work is highly detailed on the Marian motifs and theological sources.

43. Barasch, 174–5; Jansen, "Similitudo," 44–50; Mayer-Thurman, 13.

44. Goodgal, 206–25; Purtle, xv-xviii; Philippot, 75–7; John Ward, "Disguised symbolism as enactive symbolism in Jan van Eyck's paintings," *Artibus et Historiae* 15 (1994): 9–53.

45. L. Charbonneau-Lassay, *Le Bestiaire du Christ*, trans. and abridged D. M. Dooling (1940; reprint New York: Parabola Books, 1991), 258–63, 369–70; Miri Rubin, *Corpus Christi. The Eucharist in Late Medieval Culture* (Cambridge: Cambridge University Press, 1991), 310–1.

46. Sam Segal, "Die Pflanzen im Genter Altar," in *De arte et libris: Festschrift Erasmus 1934–1984* (Amsterdam: n.p., 1984), 403–20; Mirella Levi D'Ancona, *The Garden of the Renaissance: Botanical Symbolism in Italian Painting* (Florence: Leo S. Olschki Editore, 1977), 159–65; Reindert L. Falkenburg, *The Fruit of Devotion. Mysticism and the Imagery of Love in Flemish Painting of the Virgin and Child, 1450–1550,* trans. Sammy Herman (Amsterdam: John Benjamins Publishing Co., 1994), 10.

47. Paul Meyvaert, "The Medieval Monastic Garden," 25–53, and Marilyn Stokstad, "The Garden as Art," 177–85, in *Medieval Gardens,* ed. Elizabeth MacDougall, *Dumbarton Oaks Colloqium on the History of Landscape Architecture* (Washington, D.C.: Dumbarton Oaks Research Library and Collection, 1986).

48. D'Ancona, *Garden,* 190–2.

49. Mirella Levi D'Ancona, *The Iconography of the Immaculate Conception in the Middle Ages and Early Renaissance* (New York: The College Art Association, 1957), 67; Margaret Freeman, *The Unicorn Tapestries* (New York: The Metropolitan Museum of Art, 1976), 23–5, 29; Odell Shepard, *The Lore of the Unicorn* (New York: Barnes & Noble, Inc., 1967), 58, 110, 152.

50. Rüdiger Robert Beer, *Unicorns—Myth and Reality,* trans. Charles M. Stern (New York: Van Nostrand Reinhold Co., 1977), 24, 41, 72–9, 95–101; Freeman, 21–5; Malcolm South, *Mythological and Fabulous Creatures* (New York: Greenwood Press, 1987), 14–18; Shepard, 282.

51. Hirn, 466.

52. Segal, 15, 17; Freeman, 131, 143; Charbonneau-Lassay, 370; Eleanor C. Marquand, "Plant Symbolism in the Unicorn Tapestries," *Parnassus* 10 (1938): 3–8, 33, 40.

53. Segal, 403, 412; Stokstad, 179; Freeman, 148.

54. Dhanens, *Ghent Altarpiece,* 19, 56, 83–7, 97; Goodgal, 184–7, 306–19; Purtle, 5–6, 20–1; Gary M. Radke, "A Note on the Iconographical Significance of St. John the Baptist in the Ghent Altarpiece," *Marsyas* 18 (1975–76): 1–6.

55. Beer, 41; Monnas, "Contemplate," fig. 12. This relationship seems to be explicit in the textile depicting the Lamb of God and the unicorn used for a fifteenth-century chasuble.

56. Dhanens, *Ghent Altarpiece,* 83–7; Lotte Brand Philip, *The Ghent Altarpiece and the Art of Jan van Eyck* (New Jersey: Princeton University Press, 1971), 61–2, 79, 99; Falkenburg, 7–9.

57. Freeman, 132; D'Ancona, *Garden*, 323; Lawrence Naftulin, "A Note on the Iconography of the van der Paele Madonna," *Oud Holland* 86 (1971): 3–8; and John Williamson, *The Oak King, The Holly King, and The Unicorn* (New York: Harper & Row, 1986), 230.

58. Markowsky, 18–22; Roover, *Money*, 17–22.

59. Most agree there is an element of creativity, yet few comment on the specifics. Cottrell, "Birds, Beasts," 39–64, 97–145; Monnas, "Silk Textiles," 147–8, 150, 152; Monnas, "Contemplate," 112; Tietzel, 217–31, 229.

60. Marien-Dugardin, 19–20; Antonino Santangelo, *The Development of Italian Textile Design from the Twelfth to the Eighteenth Century*, trans. P. Craig (Milan: Zwemmer, 1959), CP 13 and 31.

61. Cottrell, "Birds, Beasts," 99–102; Tietzel, 229; Lisa Monnas, "Silk textiles in the paintings of Bernardo Daddi, Andrea di Cione, and their followers," *Zeitschrift für Kunstgeschichte* 53 (1990): 39–58, 44; Anne Wardwell, "The Stylistic Development of Fourteen and Fifteenth Century Italian Silk Design," *Aachener Kunstblatter* 47 (1976–77): 177–226; Klesse, 217–25.

62. Cottrell, "Birds, Beasts," 76–81; Klesse, 218; Trench, 66, 70; Wardwell, "Stylistic Development," 177–9.

63. Otto von Falke, *Kunstgeschichte der Seidenweberei* (Berlin: Wasmuth, 1913), fig. 411.

64. Perhaps the changes were an attempt to update the pattern. This practice would echo Jan van Eyck's method of duplication of the velvet patterns of the Ghent organist and Gabriel in the *Washington Annunciation*.

65. Goodgal, 133, 140; Trench, 72; James Marrow, "Symbol and Meaning in Northern European Art of the Late Middle Ages and the Early Renaissance," *Simiolus* 16 (1986): 150–69; Ward, 12.

66. Ward, 12.

CHAPTER 11

MARKED DIFFERENCE:
EARRINGS AND "THE OTHER"
IN FIFTEENTH-CENTURY FLEMISH ART

Penny Howell Jolly

While extensive scholarship already exists on Rogier van der Weyden's *Columba Altarpiece* (Munich, Alte Pinakothek), an influential work from the early 1450s originally in Cologne, two details in the triptych's central *Adoration of the Magi* (Figure 11.1) deserve further consideration.[1] Visible in the original, but difficult to see in reproductions, are two figures who wear earrings: a black man in the Magi's entourage, framed behind the stable by the backmost arched window; and a bearded man in a turban standing in the closer archway at the right. Rogier, drawing on newly developing conventions in fifteenth-century Flemish art, uses earrings to mark both figures as outsiders: "others" with regard to the Christian society of fifteenth-century Northern Europe. But rather than motivating rejection of these outsiders by viewers, earrings signal acceptance. By intensifying figures' outsider status, earrings make their eventual conversion all the more remarkable. Rogier's black attendant will soon see the Christ Child, and will acknowledge him as Son of God; the turbaned and earringed onlooker, a Jew, is this very moment "seeing the light." This new use of earrings, which develops in fifteenth-century Flanders, can be called the *even he/she* topos, meaning *even such a non-believing outsider as he/she* can experience a revelation and convert to Christianity. The power of being a Christian insider in fifteenth-century Flemish society is reinforced by heightening the otherness of the earringed outsider; Christianity's universal appeal is reconfirmed; and revelation and salvation are possible for all, because *even he/she* responds to Christianity and therefore converts.

Figure 11.1   Rogier van der Weyden, *Adoration of the Magi*. Alte Pinakothek, Munich, Germany. Copyright © Foto Marburg / Art Resource, NY.

Earrings have drawn only slight attention in the art historical literature on medieval and Renaissance art. Most notably, Diane Owen Hughes explored the function of earrings in Italian trecento, quattrocento, and early cinquecento society. She concluded that, in both life and art, earrings most commonly designated impurity and marked Jews;[2] in some regions, Jews were even required to wear them by law, a particularly offensive decree given Old Testament mandates about not cutting the body.[3] While earrings do not condemn the wearer in early works like Ambrogio Lorenzetti's 1342 *Presentation in the Temple,* where both Mary and an attendant wear earrings marking them as Jews while they attend to the mandated Hebrew purification ritual, Hughes notes slightly later images of Mary Magdalene and personifications of Vainglory with earrings where the jewelry indicates their wearers' interest in sexual and worldly matters.[4] By the fifteenth century, earrings were worn by neither the Virgin in art, nor contemporary Italian gentiles in life.

This analysis of Northern art suggests somewhat different and generally more positive functions for earrings. For example, while negative connotations continue regarding the Magdalene's wearing of earrings in texts such as Jean Michel's *Passion* play, it seems that no fifteenth-century Northern paintings depict her with earrings.[5] Earrings are, however, used in images to indicate outsiders. Studies of jewelry customs confirm what early portraits in France, Germany, and the Lowlands suggest: real men and women did not wear earrings in late-fourteenth- and fifteenth-century Northern Europe. While earrings were popular in the ancient and early medieval worlds, and remained fashionable in later centuries in the Byzantine sphere of influence, and thus in parts of both Eastern and Southern Europe, by the later Middle Ages in the North, they were curiosities.[6] Therefore their appearance in fifteenth-century imagery from this region signals their role as pictorial conventions, in this case, signifiers of difference. Only in the sixteenth century do fashions change and contemporary women begin to wear earrings, first in Italy and then, late in the century, in the North.

This study is by necessity preliminary and limited in its claims, although not its examples, to fifteenth-century Flemish art. Trying to see earrings in reproductions, while difficult, must sometimes suffice, since it is impossible to view all works in the original, especially manuscript illuminations. Most earrings discussed here were discovered after careful viewing of originals; some were visible in good quality reproductions. To date, examination of Franco-Flemish manuscript illuminations, mostly from reproductions, has revealed no earrings prior to the time of the Limbourgs' *Très Riches Heures* of 1413–16. Confirmation of this will depend upon future observations by scholars in the field, and is significant because scholars of early Netherlandish art typically turn to these illuminations when researching the origins of fifteenth-century painting conventions. Current evidence suggests earrings enter Northern art from Southern and/or Eastern Europe, appearing first in the art of the Limbourgs. They begin to interest Northern panel painters in the first half of the century, becoming more common by the last quarter of the century. In all cases, earrings mark outsiders, but this chapter will explore other connotations that develop as Flemish artists incorporate the new motif into their art.

The earliest regular appearances of earrings in Northern art are in manuscript illuminations and panels from the second decade of the fifteenth century. One clear category of earring wearers in these earliest Northern images consists of black Africans (or black Asians), where the earring emphasizes these figures' otherness. The Limbourgs depict gold loop earrings in several miniatures in their *Très Riches Heures* (Chantilly, Musée Condé), for example, on black adults and children gazing upward toward Christ in *David Sees Christ Exalted Above All Things* (f. 27v) and on several blacks in a congregation listening to an apostle preach in *David Foretells the Mission of the Apostles* (f. 28). These last Millard Meiss identifies as Indians.[7] About the same time, the Master of the Harvard Hannibal includes an earring on a black male kneeling to Hannibal's right in *Hannibal Crowned as Leader of the Carthaginians* (Harvard College Library, Richardson 32, II, f. 263). In all these, the artists' introduction of earringed blacks alongside whites stresses the universality of the scene's central event and thus its importance to members of widely differing ethnic groups, whether gathered around a vision of Christ, listening to proselytizing Apostles, or hailing the new king of the Carthaginians.

The *tituli* below the Limbourgs' illuminations confirm this reading, demonstrating the positive role these non-Europeans play with regard to Christianity.[8] The first, illustrating Psalm 8's celebration of the wonders of creation, reads, "David in the spirit sees Christ, less than the angels, ascend above every creature," and demonstrates the range of God's creation by including blacks. The second, referring to Psalm 18:4–5 ("There are no speeches nor languages, where their voices are not heard. Their sound hath gone forth into all the earth and their words unto the ends of the world"), anticipates the universality of the Apostles' message by stating "David in the spirit announces that the apostles after the ascension will spread the gospel throughout the world." For these figures, already marked by their skin color as non-European, earrings intensify their difference. Yet the earrings signal a second factor, for these persons appear within a scenario of acceptance, a setting wherein their otherness is valued and necessary to the essential meaning of the event's universality.[9]

The "double" marking of blacks as non-Europeans by addition of an earring appears most consistently and numerously in Flemish art in *Adorations of the Magi,* and again emphasizes an outsider who is accepted. According to Paul Kaplan, black attendants first appear in mid-thirteenth-century Italian *Adorations;* by the opening of the fifteenth century, some Italian and Spanish examples include earringed black attendants to all-white Magi, thus intensifying the diversity and exoticism of the peoples over whom those Magi rule and who will now accept Christ.[10] Indeed, these earliest appearances of earrings on blacks in Southern European art is not surprising, given the Mediterranean cultures' closer contacts with Africa and the East, and their continuing fashion for earrings. By contrast, French and Flemish artists rarely even include black attendants for the three Kings until the mid-fifteenth century; again, the Limbourgs' *Très Riches Heures* is exceptional, with black attendants in both the *Meeting of the Magi* (f. 51v) and the *Adoration* (f. 52).[11] Rogier's depiction of an earringed black man in the Magi's entourage in the *Columba Altarpiece* from the 1450s (Figure 11.1) is therefore noteworthy as apparently the earliest black attendant with an earring found in Flemish panel painting.[12] Rogier includes him to emphasize the universal appeal of Christ, as this distinctive outsider is about to enter the Christian fold.

The black Magus originates later than the black attendant in art, first appearing in fourteenth-century Bohemian imperial imagery stressing universal rule, and later gaining popularity in Germany; Kaplan cites Hans Multscher's 1437 *Wurzach Altarpiece* as the first incontrovertible example extant (a black attendant is also present, but neither is earringed).[13] Earrings appear soon after: perhaps the first earringed black Magus in European art is in the Bavarian Master of the *Polling Altarpiece*'s eponymous work of 1444 (Munich, Alte Pinakothek).[14] But at almost the exact same time, Dirk Bouts in his Prado *Adoration* paints his third Magus not only with African features (a broadly flattened face and tightly curled hair, and darkened although not "black" skin), but also with a gold loop in his ear.[15] This appears to be the first such Magus with earring in early Netherlandish art. The motif quickly gains popularity in Northern Europe in the second half of the century, as seen in two works closely based on Rogier's *Columba Altarpiece:* one by an anonymous Cologne painter working soon after Rogier (New York, Metropolitan Museum of Art), the other the eponymous work by the Master of the Prado *Adoration* from the 1470s (Madrid, The Prado).[16] Both retain the raised-arm posture and placement of Rogier's third Magus, but transform him into a black Magus wearing an earring. By the time Hans Memling paints two *Adorations* in the 1470s (Madrid, The Prado, ca. 1470–72 and Bruges, St. John's Hospital, 1479), both include the newly established motif, and his *Seven Joys of the Virgin* (Munich, Alte Pinakothek) of 1480 inserts multiple scenes of the Magi and their entourage journeying, including one black Magus and three black attendants, all of whom sport gold earrings. By the end of the century, the earringed black Magus is well established in Flemish art by painters such as Hugo van der Goes, Gerard David, the Masters of the St. Barbara Legend and of the *Wenemaer Triptych,* Hieronymus Bosch, Jan Provost, Joos van Cleve, Quentin Massys, and the Antwerp Mannerists. In all these, the earringed Magus demonstrates the powerfully universal appeal of Christianity.

The Italian tradition of marking Jews with earrings does inspire some following in the North, although it remains relatively rare. Once again, the Limbourgs appear to have been the first: one woman in the group kneeling behind Mary in the *Adoration of the Magi* (f. 52) displays an elaborate earring, and another accompanying Mary and Christ to the Temple for Mary's *Purification* (f. 54v) wears two pendants that dangle near her right ear lobe.[17] Functioning as female attendants during Mary's time of confinement and churching, surely both are intended to be Jews. Their jewelry signifies their difference, but probably without negative connotations.

The same pertains to earringed male onlookers present at Joseph and Mary's marriage in two different paintings of that ceremony from the last quarter of the fifteenth century by the Master of the Tiburtine Sibyl (Philadelphia Museum of Art) and the Master of the View of St. Gudule (Utrecht, Convent of St. Catherine).[18] They stand among the disappointed suitors behind Joseph, and so certainly represent Jews.

But sometimes earrings do mark evil detractors or undesirables, nonbelievers without any likelihood of conversion. Again it is the Limbourgs who first represent this probably Italian-inspired tradition in the North in their *Martyrdom of St. Mark* (f. 19v), where an earringed black man drags Mark through Alexandria.[19] This usage does not reappear consistently in early Netherlandish art, however, until the very last decades of the century, e.g., in Geertgen tot sint Jans' *Burning the Relics of John the Baptist* (Vienna, Kunsthistorisches Museum)[20] and, most notably, in the oeuvre of Hieronymus Bosch, where some of his most vehement tormentors of Christ sport rings that pierce ears and other body parts. In Bosch's early *Ecce Homo* (Frankfurt, Städelsches Kunstinstitut) of ca. 1480–85, the male detractor who prods Christ from behind while slipping his arm up Christ's cloak wears an earring.[21] So does the large-nosed man in profile near the good thief in the upper right of Bosch's much later *Christ Carrying the Cross* of ca. 1515 (Ghent, Musée des Beaux-Arts). But Bosch goes further in this late painting, using golden rings with pendant beads to pierce chins and cheeks of several grotesque detractors.[22] And while he continues the Flemish convention of depicting the black Magus with earrings in his Prado *Adoration of the Magi* from ca. 1510, the false Antichrist standing in the doorway displays this same earring—but piercing his thigh.[23] These extra face and leg rings link their wearers to Judaism, but in a blatantly negative manner: their displacement onto other body parts confirms the confused and misdirected states of these vehement nonbelievers.

Possibly the first example of a Jewish woman marked by an earring in Flemish panel painting occurs in Jacques Daret's *Nativity* (Figure 11.2; Madrid, Thyssen-Bornemisza Collection) of 1434–35, a panel that reflects a refinement of the motif's meaning that originates in the circle of Daret's teacher, Robert Campin. Campin is probably inspired by Italian tradition, but rather than suggesting rejection of the earring-wearer, as Bosch much later will do, his use emphasizes acceptance. In Daret's panel, two elaborately dressed Jewish midwives kneel, but the one to the right is distinguished as the disbelieving Salome by her posture and the beaded gold loop in her ear: the other, Zebel, remains earringless. Daret uses the earring as an intensifier, a means of demonstrating the greater distance between Salome and those others surrounding Christ who already accept Mary as his virgin mother and him as the Son of God. As the *Golden Leg-*

Figure 11.2   Jacques Daret, *Nativity.* Thyssen Bornemisza Collection. Copyright © Museo Thyssen-Bornemisza, Madrid.

end recounts the miracle, Zebel immediately recognizes Mary as a virgin who has given birth, while skeptical Salome instead reaches to examine her.[24] As punishment, her hand withers, only to be restored after an angel appears—above her in Daret's panel—directing her to touch Christ. Thus viewers witness her truly remarkable conversion, as she, the greater non-believer, becomes enlightened and accepts Christ as the Messiah. The earring heightens her initial spiritual recalcitrance, and therefore intensifies the story's miraculous outcome: her newfound belief. Salome and Zebel at first represent two categories of Jews, those who stubbornly refuse to accept Christ, and those who see the light and convert. Salome reinforces

Figure 11.3   Anon. Copy, Robert Campin, *Deposition*. National Museums and galleries on Merseyside. Walker Art Gallery, Liverpool.

and then collapses that polarity. For Daret, earrings mark the *even she* topos, in which the most adamant nonbeliever reforms and is saved. Drawing attention to this dramatic conversion, the earring does not refer to a static condition of being a Jew, as it does in the Italian tradition, but rather marks the transitional moment of reform.

This is Campin's contribution to the meaning of the earring, the *even he/she* topos. The newly expanded usage appears in earlier works linked to Campin, including his monumental *Deposition* triptych, known today through a fragment of the original (Frankfurt, Städelisches Institut, ca. 1430) and an early and highly accurate copy of it (Figure 11.3; Liverpool, Art Gallery, painted between 1448 and 1467). There an earringed man climbs the ladder to assist in the removal of Christ's body. Behind him two disbelievers gesticulate, but remain obdurate in their ignorance of the way to salvation; as Albert Châtelet notes, they turn from Christ.[25] The well-dressed, turbaned Jew on the ladder, however, is marked by a gold loop in his ear, and has reverently removed his shoes: once a pagan nonbeliever, *even he* now sees the light and accepts Christ as the Son of God. A *Crucifixion* from the second half of the fifteenth century, today in Poznan, has been identified as a copy of a lost work by Campin from ca. 1410–15.[26] Prominently seated on a white horse in the right foreground of the scene, the side traditionally reserved for Christ's tormentors, is an exotically dressed and turbaned "other": bearded, he wears a scimitar at his waist and displays a prominent earring. His pose and gesture, however, confirm his conversion: he now peers intently at Christ, his right hand raised up above his forehead. Like Salome, and like the Jew in the *Deposition,* he is depicted at the moment of revelation, as *even he* acknowledges Christ's divinity and will be saved.

Campin's *even he/she* use for the earring continues with later artists, especially in depictions of Christ's Passion. A *Crucifixion* by a follower of Simon Marmion from the 1470s (Philadelphia Museum of Art) includes an earringed black man in the most sinister location relative to the cross: to Christ's extreme left.[27] Barefoot, with long staff and shield, he wears a peculiar conical hat that further suggests his "otherness." Yet, unlike the two disputing men adjacent to him who ignore Christ, he looks up, clearly acknowledging Christ as the Savior. A similar earringed figure appears to the far right in the Master of the Tiburtine Sibyl's *Crucifixion* from the 1480s (Detroit, Institute of Art), except he is not a black gentile, but a bearded Jew who gazes at Christ and converts.[28] This becomes a popular type in German art as well, as in Derick Baegert's 1477/78 fragmentary *Crucifixion* (Madrid, Thyssen-Bornemisza Collection).[29] Not only does the once pagan male "other" staring intently at Christ from the right of the cross wear an elaborate loop-and-chain earring that calls attention to his transformation, but an earringed black woman on the other side also gazes with acceptance at Christ. The child who hangs over her shoulder has also seen the light, for he reaches his young arm toward Veronica's veil.

Other pagan figures who experience revelations bear the mark of earrings, including sibyls, the pre-Christian gentiles who, through divine revelation, anticipated the coming of Christ. Jan van Eyck's Erythrean Sibyl[30] from the exterior of the 1432 *Ghent Altarpiece* (Ghent, St. Bavon) and Rogier van der Weyden's Tiburtine Sibyl from his ca. 1445–50 *Bladelin Altarpiece* (Berlin, Gemäldegalerie) both wear gold loops. The woman in the entourage of St. Helena, in Simon Marmion's *Proving of the True Cross* from the 1450s or 1460s (Paris, Louvre),[31] kneeling in open-handed amazement at our far right as the power of the True Cross resurrects a corpse, wears an earring to highlight both the unlikeliness and the here-depicted actuality of her conversion following the miracle. As with most *even he/she* figures, her location in the picture's least holy position, its lower right, further emphasizes the improbability of her conversion. The Queen of Sheba, like the sibyls, is another pre-Christian who anticipates the coming of Christ; she also often wears an earring. Earrings certainly mark her otherness—while only rarely depicted as a black, she ruled over black gentiles (called Ethiopians or Nubians) in an exotic land variously described as in northeast Africa or Arabia—but also heighten her role as an Old Testament type for the Magi, and perhaps recall her revelation at the bridge.[32] This last miracle forms an earlier episode from the legend of the True Cross that also includes Helena's miracle.

In the Master of the Legend of St. Barbara's *Queen of Sheba Brings Gifts to Solomon* from about 1480 (New York, Metropolitan Museum of Art), the right wing of a triptych, both the queen and her male attendant wear earrings, and come before Solomon bearing gifts. The earrings enhance the scene's traditional function as a type for the arrival of the Magi; indeed, the triptych's central panel, an *Adoration of the Magi* (Rome, Galleria Colonna), includes an earringed, black Magus.[33] The inclusion of earrings in both the Old and New Testament narratives marks the otherness of these diverse peoples who arrive before the king, whether Solomon or Christ, and also signals the remarkableness of their acceptance of him. But the earrings on the queen may also refer to the miraculous revelation experienced by her and her entourage on their way to meet Solomon: she refused to walk on a bridge after recognizing it as the wood upon which the Savior of the World would later be crucified. Because of this, her malformed feet are cured. In the ca. 1440 *Hours of Catherine of Cleves,* an earringless Queen of Sheba fords the stream alongside the miracle-inducing wooden bridge, but one of the two female onlookers on the shore is black and wears an earring.[34] It certainly marks her exotic otherness, but perhaps further heightens her role as witness to these miraculous events.

In the Columba *Adoration of the Magi* (Figure 11.1), Rogier's inclusion of earrings on two figures reflects very recent innovations in Northern painting. For the earringed black attendant of the all-white Magi, perhaps the first extant in Flemish painting, Rogier uses a motif introduced to the North by the Limbourgs; it functions visually to suggest the universality of Christianity's appeal, even to the most distant outsiders. But Rogier also uses a modification of earring symbolism for the turbaned man in his *Adoration* that developed in the shop of his teacher Campin: the *even he/she* topos. Alfred Acres identifies him as a Jew—his turban and yellow garment suggest this—but as Acres also remarks, with the help of the adjacent figure who touches his arm and points, the Jew recognizes the Messiah.[35] By marking this Jew with an earring and depicting him at the very moment of his enlightenment, Rogier heightens the drama of conversion. The earring signals the deep doubt of his initial response; simultaneously it draws attention to his remarkable conversion. Thus the message to fifteenth-century viewers was one of confidence in the possibility of salvation: if even outsiders such as Jews and Africans can recognize the Messiah and be saved, hope exists for all.

Notes

This work was funded in part by a Skidmore College Faculty Research Grant.

1. Alfred Acres, "The Columba Altarpiece and the Time of the World," *Art Bulletin* 80 (1998): 422, offers an excellent overview of the scholarly literature, and includes a color detail (his fig. 21) of the relevant part of the panel.

2. "Earrings for Circumcision: Distinction and Purification in the Italian Renaissance City," in *Persons in Groups,* ed. Richard C. Trexler (Binghamton, New York: Medieval and Renaissance Texts and Studies, 1985): 155–77, and expanded as "Distinguishing Signs: Ear-Rings, Jews and Franciscan Rhetoric in the Italian Renaissance City," *Past and Present* 112 (1986): 3–59.

3. Deut. 14:1, Lev. 19:28 and 21:5.

4. Hughes, "Distinguishing Signs," 31–32.

5. See Gustave Cohen, *Le Livre de Conduite du Régisseur et le Compte des Dépenses pour le Mystère de la Passion* (Paris: Librairie Ancienne Honoré Champion), 182.

6. Ronald Lightbown, *Mediaeval European Jewellry* (London: Victoria and Albert Museum, 1992), 293–4, notes the surprise of a Burgundian traveler in Constantinople in 1432, who observed the Empress wearing earrings, and cites the 1352 French royal accounts' listing of earrings, although those were for the Dauphin's fool. Otherwise, earrings "seem scarcely to have been worn at all" in Northern Europe in this time. See also Joan Evans, *A History of Jewellery, 1100–1870* (Boston: Boston Book and Art, 1970), 47.

7. In *French Painting in the Time of Jean de Berry: The Limbourgs and Their Contemporaries,* 2 vols. (New York: Braziller, 1974), I, 219, where Meiss discusses European confusions regarding the land of origin of these and other blacks. Excellent color reproductions of both are in Raymond Cazelles and Johannes Rathofer, *Illuminations of Heaven and Earth: The Glories of the Très Riches Heures du Duc de Berry* (New York: Harry N. Abrams, 1988), 76.

8. See Margaret M. Manion, "Psalter Illustration in the *Très Riches Heures* of Jean de Berry," *Gesta* 34/2 (1995): 149–50, regarding these quotes and illuminations.

9. The German artist Stephan Lochner uses earrings similarly in his *Martyrdom of St. Andrew* (ca. 1435–40; Frankfurt, Städelisches Institut). There earrings on two Caucasian women within a crowd assist in marking the

populace as inclusive of truly diverse foreigners who recognize the truth of the Christian God. The *Golden Legend* describes Andrew's martyrdom in Achaea, when—still attached to the cross—he preached to twenty thousand people and swayed them to his side. Lochner encapsulates this crowd into nine witnesses, depicting the listeners' diversity through inclusion of both older and younger men and women, a child, dark and fair complexioned figures, and two earringed women.

10. His *Rise of the Black Magus in Western Art* (Ann Arbor, 1985), 7–17, explores the development of the black attendant, but without mention of earrings. Early examples of earringed attendants include the Aragonese altarpiece by Nicolàs Solana (1401–1407, Madrid, Instituto Valencia), ibid., fig. 18; and Jacobello del Fiore's early fifteenth-century *Adoration* (Stockholm, Museum), illustrated in Liana Castelfranchi-Vegas, *Il Gotico Internazionale in Italia* (Rome: Editori Riuniti, 1966), pl. 47.

11. Illustrated in Cazelles and Rathofer, *Illuminations of Heaven and Earth,* 89–90, 92–3. Kaplan, *Rise of the Black Magus,* 15, identifies the African in the *Adoration* as "probably" the first Northern example of a black attendant, but this figure appears not to wear earrings, although a midwife there does. Italian artists included black attendants much earlier than Northern; ibid., 7, lists Nicola Pisano's *Adoration* from his Siena Cathedral Pulpit of 1266–68 as the first, but no earrings are present.

12. Ibid., 16, without discussing earrings, confirms the rarity of black attendants in French and Flemish art of the time, and believes Rogier was influenced by Italian art. Joos van Gent's *Adoration of the Magi* (New York, Metropolitan Museum of Art), of ca. 1465, includes another early Flemish example of an earringed black attendant.

13. Ibid., 87–9. The black Magus in a Bohemian fresco at the Monastery at Emmaus from ca. 1360, Kaplan's earliest extant example, already shows this third King sporting an earring, but Kaplan notes the figure may be repainted. Thus, the earring may not be original.

14. Ibid., fig. 61.

15. Illustrated in Albert Châtelet, *Early Dutch Painting,* trans. C. Brown and A. Turner (Secaucus: Wellfleet Press, 1980), 77.

16. The first is illustrated in Charles Sterling et al., *Robert Lehman Collection, II: Fifteenth- to Eighteenth-Century European Paintings* (New York: Metropolitan Museum of Art, 1988), 34–6; the second in John Hand and Martha Wolff, *Early Netherlandish Painting* (Washington: National Gallery of Art, 1986), 156.

17. Illustrated in Cazelles and Rathofer, *Illuminations of Heaven and Earth,* 93 and 95.

18. The former is illustrated in Châtelet, *Early Dutch Painting,* 141; the latter in *Anonieme Vlaamse Primitieven* (Bruges: Stad Brugge, 1969), 118.

19. Illustrated in Cazelles and Rathofer, *Illuminations of Heaven and Earth,* 67.

20. Painted after 1484; illustrated in Châtelet, *Early Dutch Painting,* 105. Geertgen offers a highly unusual back view of a black man who stands in the entourage of Julian the Apostate, wearing two prominent earrings.

21. Illustrated in Charles de Tolnay, *Hieronymus Bosch* (New York: Reynal and Co., 1966), 82 and 85.

22. Illustrated in ibid., 310–11. St. Veronica, in the lower left corner, wears a headdress with hanging strings and beads, one of which hangs near her ear and thus looks similar to an earring, although it is not specifically one. Her delicate pendants and idealized face contrast markedly with those of Christ's tormentors.

23. Illustrated in ibid., 297–8.

24. *The Golden Legend of Jacobus de Voragine,* trans. Granger Ryan and Helmut Ripperger (New York: Arno Press, 1969), 48.

25. In *Robert Campin; Le Maître de Flémalle* (Antwerp: Fonds Mercator, 1996), 84, where he identifies the man high on the ladder as Nicodemus and— without reference to his earring—the man at the bottom of the ladder (excellent color detail, 79) as a converted Jew. I think it possible that Nicodemus is the converted Jew at the bottom of the ladder, as he is much better dressed than his assistant above. Scholars accept the Liverpool copy as a highly accurate one, based on comparison of it and the extant fragment of Campin's original.

26. Ibid., 63 and 309.

27. Illustrated in Thomas Kren, ed., *Margaret of York, Simon Marmion, and The Visions of Tondal* (Malibu: The J. Paul Getty Museum, 1992), fig. 251, and in *The John G. Johnson Collection: Catalogue of Flemish and Dutch Paintings* (Philadelphia: John G. Johnson Collection, 1972), no. 318.

28. Illustrated in Châtelet, *Early Dutch Painting,* 140.

29. Illustrated in Isolde Lübbeke, *Thyssen-Bornemisza Collection: Early German Painting 1350–1550,* trans. M. T. Will (London: Sotheby's Publications, 1991), no. 26–7.

30. Jan's is typically but erroneously identified as the Cumaean Sibyl, due to the reversed labels on the altarpiece's frame for the two sibyls. The texts they hold confirm their correct identities.

31. Illustrated in Kren, ed., *Margaret of York,* 177, fig. 130.

32. Kaplan, *Rise of the Black Magus,* 9 and 37–41, discusses the tradition of the black Sheba, and her typological role regarding the Magi. He cites Nicolas of Verdun's Klosterneuburg Altarpiece of 1181 as the earliest extant example of a black Sheba; already there she wears an earring.

33. The altarpiece is illustrated in *From Van Eyck to Bruegel: Early Netherlandish Painting in The Metropolitan Museum of Art,* ed. Maryan Ainsworth and Keith Christiansen (New York, Metropolitan Museum of Art, 1998), 121–2. It is a traditional pairing of Old and New Testaments scenes, found, for example, in the *Biblia Pauperum.*

34. New York, Pierpont Morgan Library, M. 917, p. 109; illustrated and discussed in John Plummer, *Hours of Catherine of Cleves* (New York: George Braziller, 1966), no. 85.

35. "The Columba Altarpiece," 444.

CHAPTER 12

THE MARGARET FITZGERALD TOMB EFFIGY: A LATE MEDIEVAL HEADDRESS AND GOWN IN ST. CANICE'S CATHEDRAL, KILKENNY

Elizabeth Wincott Heckett

Introduction

The genesis of this research was a working brief prepared for the National Museum of Ireland, Dublin for a reconstruction of the dress of Margaret Fitzgerald, wife of the eighth Earl of Ormond as shown in their effigial sculpture in St. Canice's Cathedral, Kilkenny, Ireland. (Figure 12.1) This city is the ancestral seat of the Earls of Ormond and is still dominated by their castle. The reason for the reconstruction is the permanent Medieval Ireland 1150–1550 exhibition installed at the National Museum in Dublin in the fall of 2001. A reconstruction of the Margaret Fitzgerald costume, together with another of the suit of armor worn by her husband, Piers Butler, forms part of the exhibition. In Irish history the end of the medieval period is deemed to be 1550 A.D. and so the Butler double effigy falls in its last part. This research was an exercise in deduction, a kind of detective work linked to the hard facts about specific textile finds established by archaeological excavations and by evidence from historical sources.

The Context

Margaret Fitzgerald was the wife of Piers Butler, Earl of Ossory and eighth Earl of Ormond, and the daughter of the Earl of Kildare, a near neighbor of the Ormond family. The Fitzgeralds and the Butlers were among the most powerful dynasties of the time. Piers and Margaret Butler were a colorful couple, ambitious and strong-willed. He was known as Ruadh (red

Figure 12.1   Double tomb effigy of Piers Butler and Margaret Fitzgerald, St. Canice's Cathedral, Kilkenny. Plate 158 from *Irish Medieval Figure Sculpture,* by John Hunt. (The Irish Picture Library, Dublin.)

or red-headed), was born into a minor branch of the Butler family but determined to become Earl and achieve power and wealth. His wife seems to have been equally strong-minded; it is recounted that at one time they were hiding from enemies in the woods. Margaret was pregnant and demanded that her husband go and find her some good wine because "shee was not able any longer to endure so streight (austere) a life." He did indeed take his life in his hands to procure a cask of wine for her.[1] Piers Butler died in 1539 A.D. and Margaret Fitzgerald in 1542 A.D. so it is possible that she commissioned the double tomb sculpture after her husband's death, or indeed that it was commissioned prior to that date. It has been suggested that could have been as early as between 1515 and 1527.[2] Piers Butler had achieved his life-long ambition to become Earl of Ormond only 18 months before he died and so the tomb sculptures may well represent an affirmation of his enhanced standing. It seems that the costume the Countess is wearing must embody the same ceremonial meaning to their contemporaries as does his suit of armor.

There are quite a number of similar though not identical double or single effigies with the men wearing armor and the women wearing voluminous gowns within a hundred year time span in Ireland, with the majority in the east of the island.[3] The form of dress of the Fitzgerald effigy is archaic, probably dating from the late fourteenth century, but with developments seemingly specific to Ireland.

In the late fifteenth and early sixteenth centuries fashion was indeed important but in different parts of Europe national trends had become specific rather than generalized. Wearing the dress of your own land showed your loyalty, and adopting other styles could be dangerous.[4] In Ireland the impetus toward this individual insularity was strengthened by the fact that the country was being ruled from England, against its wishes, by Henry VIII and his ministers in London. To the normal dictatorial tendencies of medieval rulers to their subjects on what they should and should not wear was added the desire of the Tudors to eradicate indigenous styles of dress in Ireland, which (to their eyes) only encouraged ideas of independence. So the Fitzgerald dress may represent the deeply held conservatism of a member of the aristocracy, a conscious choice that underlined her independence of thought, and an affirmation of her place in society.

At the time of her death, Margaret Fitzgerald, Countess of Ormond was a mature woman with a grown-up family. Indeed, one of her sons lived for many years at the court of Henry VIII, and the Ormond family was connected by marriage with that of the Boleyns. This son James was at one time thought of as a husband for his cousin Anne Boleyn but unfortunately for her she went on to better things. (She, of course, married Henry VIII and was executed by him.)[5] It is clear that the Butlers would have known

all about the latest fashions but in this pair of effigies—their final statement—they sidestepped them completely. They were in tune with the cosmopolitan tendencies of Renaissance Europe. For example, they were
aware of developments in education and founded a grammar school in
Kilkenny.[6] About 1525 A.D. to encourage local industries they brought
over Flemish weavers to Kilkenny Castle to make "diapers, tappestries,
Turkey carpets, cushions, and other like works."[7]

The tomb sculpture shows a heavy dress or gown, belted just below the
bust, and an elaborate stiffened headdress. It is an open question how frequently this type of ceremonial dress would have been worn. While there
are obvious limitations on the information available from a sculpture, certain deductions can be made from careful observation as to the construction of the robe and headdress. The following describes this observation:
The Headdress is two-horned, and is a heavy solid piece that presumably
needed buckram and/or felt stuffing to maintain its shape. (Figure 12.2) It
is made up of four elements: an under-cap or band, two matching side
panels on the horns, a central panel linking the two horns, a veil suspended
from the top of the horns. The sculpture shows a band of material across
the forehead below the headdress that must represent either a band tied
around the head, or the *under-cap*. This would have been necessary to provide a foundation to support the headdress. The cap/band is likely to have
been made from plain weave (tabby) fine linen.

The *horns* are a style of headdress that was widely worn in Europe in
the fifteenth century. It had obviously persisted in Ireland, although with
certain differences and development. Drawing from earlier examples it
seems that the side parts of the horns are made of thick netting, either from
silk or gold metal threads. The last were made by thin strips of metal being
twisted around a core of silk or linen thread; by the early sixteenth century the metal may just have been drawn and then beaten flat. (The earlier method was to beat out the metal very finely and then cut it into
narrow strips.) The netting would be stretched over a panel of cloth, probably silk.

In the earlier types of headdress, the hair was braided over the ears and
enclosed in decorative cauls of bejeweled gold netting that later developed
into the side panels. These reticulated cases were known as bosses, or tem-
plers, and might have been lined with silk.[8] The two horns and the interior structure would be made first and then the central panel applied
afterward. This can be seen from where the edging for the horns disappears
under the central panel. A parallel for the construction of the headdress may
be drawn with medieval bishops' miters that until the end of the twelfth
century were worn with the horns at each side of the head rather than at
the front and back. A late fourteenth-century French example in silk is
decorated with raised and couched embroidery showing biblical scenes.[9]

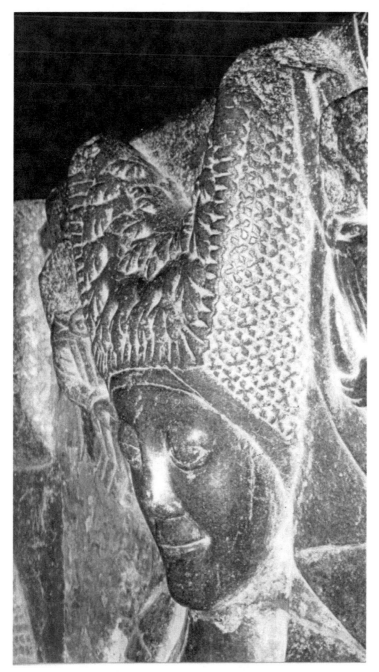

Figure 12.2   Detail of the Fitzgerald headdress, St. Canice's Cathedral. Kilkenny. Photograph by Elizabeth Wincott Heckett.

It seems that both secular and ecclesiastical headdresses may have been stiffened with a lining of buckram, nowadays a coarse cotton fabric heavily sized with glue and still used for such purposes. We know from English Royal Great Wardrobe Accounts that it was used in this way in the fourteenth century. It is recorded in the accounts for Christmas masquerades in 1345 A.D. that 15 pieces of *bokeram* were among the items bought for the construction of fantastic headdresses.[10]

The *central panel* shows raised or embossed decoration on the surface of the cloth. Earlier fifteenth-century headdresses were embroidered, with short transparent veils attached to the horns; and the central roll between them is described and illustrated by Peacock as beaded and embroidered.[11] This may well have developed into the Irish variant of the roll tilting smoothly between the two horns.

One way of producing this effect would be with embroidery. Irishwomen's clothes at this time were certainly embroidered and decorated since there was a royal decree prohibiting items that were "imbroydered or garnished with silke, or courched [couched] ne layd with usker "(Gaelic *usgar*, a jewel or ornament).[12] Such legislation is a sure sign that the behavior that is forbidden is in fact being carried out! There are no surviving examples of Irish embroidery of the period, but perhaps a connection may be made with Icelandic practices, since they stem from Norwegian and Celtic settlers who established themselves there at about the same time the Norse came to Ireland.[13]

In the circumstances it seemed legitimate to explore some examples of Icelandic stitching that would give a raised effect, and also other less specific couching stitches.[14] An example of seventeenth century Spanish embroidery of stylized floral scrollwork is also a useful comparison.[15] French raised and padded *broderie en relief,* and similar work—even using an underlay of wood to create an embossed effect—is known from Austria and Germany.[16] The raised effect so clearly seen on the sculpture may have been achieved with unspun wool or cotton pushed in at the back of the work after the decorative stitching was in place. This technique was used in the late fourteenth century Guicciardini quilt from Sicily.[17] Another type of raised work was formed with a padding of linen threads, as seen in a couched gold and silver embroidery on a German hat of about 1600 A.D. This hat has a moulded thick felt base that had been stiffened with size, the stiffening agent made from protein glue.[18] It may be that the Fitzgerald headdress was made up with a similar foundation.

It is also possible that a luxury velvet made from cut and uncut pile of differing heights, like one held in the Abegg-Stiftung, Riggisberg, Berne, Switzerland, could create a similar effect. This is a piece from the early sixteenth century with large motifs and rich coloring.[19] Something similar to this High Renaissance brocaded velvet could perhaps have been appliquéd onto the cloth to produce the effect shown on the headdress. The leafy motif has some thing in common with the acanthus-type leaves on its central panel. However, the way the pattern fits in so well with the shape of the headdress strongly suggests that the decoration was specifically designed to fit the panel, and so it is more likely to have been embroidery.

Two little angels at the top of the horns are shown holding *a fine veil* in place that falls to the base of the throat. Contemporary English headdresses and that of Anne of Cleves, the German princess (1512–48 A.D.) who married Henry VIII show the use of very fine, almost transparent linen.[20] It seems likely that Margaret Fitzgerald's veil would have been made of such lawn. The veil may be extended and so kept in place by fine wire. Although the style of the costume is clearly archaic, the actual materials used were most probably contemporary, so that portraits of the time can give useful clues as to the cloth and decorative trimmings available. It should be noted that Anne of Cleves is wearing a hair net made from thick gold-colored silk strands whose ends seem to be knotted to the left-hand side of her head. The portrait gives a good idea of the type of silk netting used at the time. Perhaps it was this sort of gold thread that was used for the side panels of the headdress. Also to be noted is the lavish use of pearls on the cauls over her (braided?) hair. This German head covering is very different to those worn at the English Court, and illustrates the point that regional styles could be quite specific.

The *gown* derives from the *houpplandes* of the last quarter of the fourteenth century that continued to be worn in many variants in the fifteenth century. One of the accompanying headdress styles worn with this gown was the earlier version of the Fitzgerald head covering. Some earlier gowns had "bagpipe" "poke," or "pudding" sleeves. Many had high waists, were belted below the bust, some with belts either embroidered or made of leather. The gowns very often fell in regular ample folds. The women shown from earlier times wore under-gowns with long tight sleeves, but in many cases these appear to be buttoned from cuff to elbow. The Fitzgerald effigy displays most of these characteristics, as can be seen in the accompanying illustration, so that the gown falls from a gentle v-shaped collar band in full, regular folds to the ground, with additional material inset below the knees. The sleeves are very wide, and gathered in above the wrist, showing the lower sleeves of the under-dress. The gown is belted under the bust with a highly decorated long belt. The sleeves of the under-gown are fastened with laces not buttons. (Figure 12.3.)

The cloth for the gown may very well have been of fine wool such as an English broadcloth which by the sixteenth century was held in high esteem throughout Europe. Although there was a tenacious retention of the

Figure 12.3   Detail of the Fitzgerald gown, St. Canice's Cathedral, Kilkenny.   Photograph by Elizabeth Wincott Heckett.

traditional styles of dress in Ireland, it seems that the cloth used may be for-eign. In the *Life of the Earl of Kildare* there is a letter of July 1539 A.D. "that Art Oge O'Toole had sent to Gerald" (son of the late Earl of Kildare) "be-fore Christmas a saffron shirte dressed with silke, and a mantell of English cloth fringed with silke."[21] The mantle with fringes was a well-established and indeed proscribed Irish fashion, and yellow saffron dye had equally long associations with Irish dress. So we can see that Irish aristocrats were wearing English cloth, whether it may have been made up in the Irish or English fashion. Satin and velvet were also worn by the nobility, as a letter from St. Leger to Henry VIII reports that the O'Donnell in 1541 A.D. was wearing crimson and black velvet, and crimson satin.[22] There does not seem to be any suggestion of velvet in the way the robe is sculpted, and a conservative noble lady of Margaret Fitzgerald's rank and age could well have chosen the very expensive English broadcloth. This would have been napped (teaseled or fulled to raise the fibers) and close sheared as many as three times to produce a fine, smooth cloth in which the weave would be invisible.

Such superfine cloth can be seen being worn by the young woman in the Arnolfini wedding portrait (1435 A.D.) by Jan van Eyck. Her dress shows off the quality and draping characteristics of this fine wool cloth.[23] There is a rare survival of samples of fifteenth century English broadcloth attached to a merchant's order book from southern France which provides a "certified" example that can be used for comparison with excavated pieces from medieval sites. These samples are said to be of "good, middle quality," but not of the first rank, since the cloth was destined for "brides of well-to-do peasants and townsfolk."[24]

The collar or neckband could have been made from a plain velvet to give some contrast to the main cloth. Catherine of Aragon (1485–1536 A.D.) first wife of Henry VIII, was a princess with strong religious convic-tions from the conservative royal house of Spain. Her portrait, painted when she was about 45 years old, shows her wearing a gown of black vel-vet with over-sleeves of dark brown fine wool, seemingly of broadcloth.[25] This may suggest the type and quality of cloth likely to be chosen by the Countess of Ormond. Excavations in the old quarter of Dublin have un-covered remains of costly silk velvet, both patterned and plain (most likely from Italy), silk and fine quality wool cloth dating to the last years of the sixteenth and early seventeenth centuries. They seem to have come from a tailor's workshop and, although later than the period in question, show that high-quality goods were indeed imported into Ireland.[26]

The belt, fastened under the bust with a buckle, falls in a long pendant down the front of the gown. It is heavily ornamented, and may be either of embroidered cloth, or embossed or decorated leather.

The Construction of the Gown

An assessment of the possible construction of the gown also had to be made. The period in question is too early to be able to take advantage of the type of analytical work so painstakingly and brilliantly carried out by Janet Arnold on dress of the later sixteenth and early seventeenth century. In her studies we also have the benefit of some of the first printed tailors' pattern books.[27] One useful source is Tarrant's *The Development of Costume,* which includes the pattern of the wedding dress of Maria of Hungary that dates to 1525.[28] The pattern shows that the skirt was made up from a full circle of cloth, with lengths of cloth having been sewn together to make up the desired fullness. The skirt falls in deep folds from the natural waist-line. Perhaps an even wider and deeper circle was the pattern for the Irish gown, allowing its longer line with the folds falling from the shoulders.

Another method that may have been used to make the gown was known in Ireland. This dressmaking technique is shown in the Shinrone dress, Co. Tipperary[29] and the Emlagh, Co. Kerry dress[30] that are dated to the late-sixteenth/early-seventeenth centuries. Both skirts have well-defined folds that were made by stitching in individual gores or panels.

The skirt of the Shinrone dress has been well analyzed and described by Dunlevy as being almost full-length and being made up of 23 triangu-lar pieces of cloth. These gores are sewn together so that the skirt measures 22 and a half feet (6.86 m) at the base. The interior vertical welts were stitched in a rough tacking stitch at regular intervals to make up the folds. The panels were sewn with thick wool thread. Each gore had four folds, one hiding the seam joining them together. This creates a skirt of 92 folds each about two and a half inches (7 cm) wide at the bottom and two inches (5 cm) at the waist. Since the weight of this skirt is considerable, the maker of the dress had to strengthen the lower part of the bodice, from bust to waist with a double layer of fabric.[31]

Shee and O'Kelly note that the Emlagh skirt is made up of 12 rectan-gular strips of four inches (10 cm) wide cloth. Each strip is then sewn with a running stitch into 13 narrow, very regularly quilted, vertical ribs. None of the lower edge of the skirt survives so the authors comment that orig-inally it may have been longer and perhaps worn in a different way.[32]

These two actual examples of careful pleating are backed up by Luke Gernon's 1620 A.D. description of how Irishwomen's skirts were made: "The skyrt is a piece of rare artifice. At every breadth of three fingers they sew it quite through with a welte, so that it seemeth so many lystes" (selvedges) "putt together. That they do for strength."[33]

The *under-dress* or *smock* would have been made of linen. The Irish *léine* or undershirt is described as using many yards of linen in the making. A fifteenth-century insular painted wood carving of God the Father from Fethard, Co. Tipperary shows clearly that the undershirt under the mantle is voluminous indeed.[34] The only parts that show on the Fitzgerald effigy are the laced sleeves already noted, which extend below the gown sleeves. The laces would have been tablet-woven (card-woven), probably of silk. This is the same technique that is now industrialized and machine-made for shoelaces. The laces worn by Margaret Fitzgerald are tubular rather than flat-weave. These again are a conservative choice for the period, since buttons had become very fashionable. An interesting corollary of the choice of laces at the wrist is that it presupposes that an attendant must be available to tie up the laces, since it is not a one-handed operation.

In the early years of the sixteenth century in Ireland the most popular imported *dyes* and *colors* were "orchell" a purple dye obtained from the lichen, *ocrolechia tartarea* and saffron, a yellow dye from the stamens of the crocus plant.[35] The first is also used in conjunction with madder to obtain a deep black, as well as on its own. Saffron has long been associated with the Irish shirt, the *léine* and it may be that a pale yellow from saffron was used for the under-dress/smock, the cuffs of which are shown below those of the gown. However, it may be that the linen veil and under-cap of the headdress were most probably white, if one follows the evidence of examples from many other countries.

Kermes, a very expensive scarlet red dye was also imported and has been found in a shaggy pile textile from Drogheda dating to ca. 1523 A.D.[36] Woad and madder were also readily available, so blues, reds, and purples were used.

It is suggested that purple be considered as the color for the gown. We have noted above that Catherine of Aragon seems to have brown fine wool over sleeves on her black gown. As a conservative and mature Spanish princess who firmly believed herself to be Queen of England, she is dressed in somber colors but in very fine quality cloth. This may point to similar choices for Margaret Fitzgerald. The *Shoes* of the sculpture are visible only by the foreparts. They appear to be plain (leather?) with slightly pointed toes.

Conclusion

The persistence of dress types over a long period of time in Ireland is not limited to the *houpplande*-derived gown worn by Margaret Fitzgerald. It has been noted and discussed in relation to, for example, the male doublet of fourteenth-century type that retained a gathered peplum, and the male and female hanging sleeves also of the fourteenth century, which all continue to be found in Irish dress until the late sixteenth century.[37] The Irish shaggy pile cloak had its antecedents in late antique times in continental Europe. Pieces of pile cloth have been found in Viking contexts in Ireland,[38] and a shaggy pile textile known as the Mantle of St. Brigid, perhaps of the eleventh century, that was held in Bruges, Belgium.[39]

On the question of whether the Ormond costumes would in truth have been worn at this late time, John Hunt's definitive discussion of Irish armor in the late Middle Ages should be noted. He is quite specific that the suits of armor shown on the figure of Piers Butler and other similar effigies are factual representations, and indeed that individual pieces may have survived over a long period of time.[40] It is perhaps less likely that the female gowns would be actual physical survivals although it is of course well established that dress of very costly cloth was given and bequeathed between generations, and so may have survived for longer periods than in modern society.

In conclusion, the costume shown in the sculpture illustrates the importance of dress in late medieval times as a marker of social identity. Then the clothes you wore defined your position in society; for example, for many people clothes were a large part of your wages and so were chosen not by you but by your employer. It is difficult for us nowadays, with our emphasis on individual choice, to think ourselves into that mind-set. When we look closely at those effigies in St. Canice's Cathedral we can catch a glimpse of that earlier world, where for contemporary observers the message was in the costume and not in other aspects of the human personality.

Notes

I am grateful to Ragnall ÓFloinn and the National Museum of Ireland for encouraging me to develop the working brief prepared for the Medieval Ireland 1150–1550 exhibition into this chapter. I would like to thank the Dean and Chapter of St. Canice's Cathedral, Kilkenny for their assistance in my research and photography.

1. Adrian Empey, "From rags to riches: Piers Butler, eighth Earl of Ormond, 1515–39," in *Journal of the Butler Society* 2, 3 (1983/84), 299–314: 306.

2. Edwin B. Rae, "Irish sepulchral monuments of the later Middle Ages," *Journal of the Royal Society of Antiquaries of Ireland,* 100 and 101 (1970–71), 1–39: 33.

3. John Hunt, *Irish Medieval Figure Sculpture 1200–1600,* vols.1 and 2 (Dublin and London: Irish University Press/Sotheby Parke Bernet 1974),Volume 1 (text) Period II 1450–1570, 61–5, 186 and Period II Civilian Ladies, 88–91, 187.Volume II (photographs) Period II: Knights including Double Effigies and Period II: Civilian Ladies.

4. Naomi Tarrant, *The Development of Costume* (London: National Museums of Scotland and Routledge, 1994), 54.

5. Adrian Empey, 310.

6. Adrian Empey, 299–301.

7. Ada Longfield, *Anglo-Irish Trade in the 16th Century* (London: Routledge and Sons, 1929), 81. Hist. Mss. Comm. App. To second Rep., 224–5 (MSS. Of O'Conor Don).

8. Joan Nunn, *Fashion in Costume 1200–2000* (London: Herbert Press, 2000), 26.

9. François Boucher, *Histoire du Costume en Occident de l'Antiquité à nos Jours* (Paris: Flammarion, 1965),186. The miter is in the Musée Cluny, Paris, and is believed to be from La Sainte Chapelle, Paris.

10. Stella Mary Newton, *Fashion in the Age of the Black Prince: a study of the years 1340–1365* (London: Boydell Press and Rowman & Littlefield, 1980), 77.

11. John Peacock, *Costume 1066–1990* (London: Thames and Hudson, 1994), 25 and 28.

12. H. F. McClintock, *Old Irish & Highland Dress* (Dundalk: W. Tempest, Dundalgan Press, 1943), 66.

13. Lanto Synge remarks that Viking and Celtic settlers in Iceland were skilled embroiderers, continuing their art to the mid-sixteenth century. Lanto Synge, *Antique Needlework* (Poole: Blandford Press, 1982), 21. Viking incomers were well settled into Ireland by the tenth century.

14. Elsa Gudjónsson, *Traditional Icelandic Embroidery* (Reykjavík: Iceland Review, 1982), 16. See also Mary Thomas, *Dictionary of Embroidery Stitches* (London: Hodder and Stoughton, 1981), 54–7.

15. Judy Juracek, *Soft surfaces—visual research for artists, architects and designers,* (London: Thames and Hudson, 1999), 141.

16. Synge, 1982, 18.

17. Kay Staniland, *Medieval Craftsmen: Embroiderers,* (London: British Museum Press, 1997), 38.

18. Janet Arnold, *Patterns of fashion,* (New York: Macmillan, 1987), 93.

19. This piece is held in the collection of the Abegg-Stiftung, Riggisberg, Switzerland, and was included in the 1996 summer exhibition of *Velvets of the West—from the Gothic to Napoleon III.*

20. The miniature of Anne of Cleves (1515–57) by Hans Holbein (1497–1543) is in the Victoria and Albert Museum, London (P153–1910).

21. This is from a letter from Allen to Thomas Cromwell, minister of Henry VIII, as quoted from the Life of the Earl of Kildare in manuscript notes by Lord Dillon held by the Royal Society of Antiquaries of Ireland, in J. H. McClintock, 66.

22. J. H. McClintock, 66.

23. The portrait of the Arnolfinis by Jean van Eyck (1435) is in the National Gallery, London. The young woman is wearing a *houpplande* gown of fine green wool lined with fur. It is belted under the bust so that small regular folds are held in place above and below the belt.

24. Philippe Wolff, "Three samples of English fifteenth-century cloth," in *Cloth and Clothing in Medieval Europe,* ed. N. B. Harte and K. G. Ponting, (London: Heinemann and Pasold Research Fund, 1983), 174–5.

25. The miniature of Catharine of Aragon, 1485–1536, by an unknown artist ca. 1530, is in the National Portrait Gallery, London (no.163).
26. Elizabeth Wincott Heckett, "'The apparel oft proclaims the man'—late sixteenth and early seventeenth century textiles from Bridge Street Upper, Dublin" in the forthcoming proceedings of 7th NESAT Symposium, Royal Museum of Scotland, Edinburgh 1999.
27. Janet Arnold, 4–10.
28. Naomi Tarrant, 55.
29. Mairead Dunlevy, *Dress in Ireland,* (London: Batsford, 1989), 48.
30. Elizabeth A. Shee and Michael. J. O'Kelly, "A Clothed Burial from Emlagh near Dingle, Co. Kerry" *Journal of the Cork Historical and Archaeological Society,* 71, 213–4 (1966): 81–91, 81–3.
31. Dunlevy, 48.
32. Shee and O'Kelly, 81–3.
33. McClintock, 95.
34. Fifteenth/early-sixteenth century Irish painted wood carving of God the Father, Fethard, Co. Tipperary, now in the National Museum of Ireland, Collins Barracks, Dublin.
35. Longfield, 180–1.
36. Elizabeth Wincott Heckett, "An Irish 'Shaggy Pile' Fabric of the 16th Century—an Insular Survival?" in *Archaeological Textiles in Northern Europe, Tidens Tand nr. 5,* report from the 4th NESAT symposium, ed. L. Bender Jørgensen and E. Munksgaard, Copenhagen (Copenhagen: Den Kongelige Danske Kunstakademi, 1992), 58–68.
37. Plate II in John Derricke's *The Image of Irelande* (1581) shows on the left of the plate a man whose doublet has hanging sleeves and tightly gathered pleats at the waist (Belfast: Blackstaff Press edition, 1985).
38. Frances Pritchard, "Aspects of the Wool Textiles from Viking Age Dublin" in *Archaeological Textiles in Northern Europe, Tidens Tand nr. 5,* report from the 4th NESAT symposium, Copenhagen, ed. L. Bender Jørgensen and E. Munksgaard, (Copenhagen: Den Kongelige Danske Kunstakademi, 1992), 95–8.
39. H. F. McClintock, "The Mantle of St. Brigid," *Journal of the Royal Society of Antiquaries of Ireland* 66, 1 (1936): 32–40.
40. John Hunt, vol.1, p. 63.

CHAPTER 13

"AS PROUD AS A DOG IN A DOUBLET":
THE IMPORTANCE OF CLOTHING
IN *THE SHOEMAKER'S HOLIDAY*

Linda Anderson

Although Thomas Dekker's *The Shoemaker's Holiday* (1599) is an early modern play, its source materials and setting are late medieval. It is a play that looks back to the feudal world of personal service to king and master, and to the camaraderie of working men within their craft guilds.[1] The editor of the New Mermaid edition of the play notes its " . . . pervasive imagery of food."[2] Oddly enough, however, in a play named for an aspect of the clothing trade, and in which that trade is central to the plot, no one seems to have commented on the pervasiveness of clothing imagery. Not only shoemaking, but all sorts of clothing is used and discussed throughout *The Shoemaker's Holiday.* Clothing is used as a status marker, an identifier, and a disguise. It is both coveted and trivialized. It is the result of honest labor and used as a bribe. It is a gift and, on at least one occasion, a kind of magic. In addition, the play's characters use clothing analogies and clothing proverbs to express a range of opinions about a variety of subjects.

Dekker's emphasis on clothing is hardly surprising, given that his principal source for the play, *The Gentle Craft,* was written by Thomas Deloney (1543?–1600), who was a clothmaker, specifically, perhaps, a silk weaver, and who dealt with the accomplishments of weavers in two of his other works, *Jack of Newbury* and *Thomas of Reading.*[3] Although Deloney seems to have turned increasingly to writing in later life, he was proud of his original occupation, declaring in his dedication to *Jack of Newbury,* "Among all manual arts used in this land, none is more famous for desert or more beneficial to the commonwealth than is the most necessary art of

clothing."[4] Nor was Deloney exaggerating: a modern commentator has
called the clothing industry England's "most basic industry," and noted that
"So important to the economy of England were the clothing industry in
general and John Winchcombe's mills in particular that in 1516 and again
in 1518 King Henry VIII visited Newbury, staying overnight in the home
of the famous clothier."[5]

In the world of the play, clothes are important in various ways. Most
simply, perhaps, as they had been throughout the Middle Ages, clothes are
a way of identifying particular kinds of people, as when the journeyman
shoemaker Firk describes his master Simon Eyre as being elected to the
office of sheriff by "a great many men in black gowns" (10.117). Firk refers
to a servant as "Bluecoat," and Eyre refers to his officers as "buff-jerkin var-
lets" (18.70, 17.23). Eyre's office of sheriff is symbolized by a gold chain
and a scarlet gown, which he refers to as "a red petticoat" (10.139, 11.12,
17.18). At the play's end, Eyre, now Lord Mayor of London, orders his
brother shoemakers, wearing the distinctively colored "satin hoods" of
their formal livery, to wait upon the king (20.1–6 and n. 4). Clothing,
therefore, does more than simply protect and adorn the body: it serves as
the most obvious sign of a person's place in the social hierarchy.

Because so much importance is placed on clothing as signifier of identity, however, it becomes the perfect vehicle for characters who wish to
disguise their true identities. In order to convince the Dutch skipper that
he has sufficient credit to buy the ship's cargo, Eyre disguises himself as an
alderman, wearing "a seal-ring . . . a guarded gown and a damask cassock"
(7.105–106).[6] It is the romantic hero Roland Lacy's disguise as a Dutch
shoemaker that allows him to dodge military service, win his own true
love Rose, escape her father's wrath when he finds them together, and
make the acquaintance of Eyre, who pleads his case to the king.[7] When
Rose's father and Lacy's uncle learn of Lacy's disguise, this knowledge only
helps convince them that the lame shoemaker Ralph and his masked wife
Jane are their disguised relatives, a mistake that gives Rose and Lacy the
time to flee to church and wed. Even the dim and ineffectual gentleman
Hammon disguises himself when he confronts Jane, although what he
hopes to gain by doing so remains obscure (12.20–21).

For most of the main characters in this play, clothing is also their means
of making a living. Not only do Eyre, Hodge, Firk, Ralph, and the disguised Roland Lacy rely on shoemaking for their livelihood, but Eyre's
wife's maid is employed to ready the thread the men need for their work
(4.34–37). Ralph's wife, Jane, is told by Eyre that when her husband goes
off to war, she must spin and card to earn her living, and she is later found
working in a sempster's shop (1.210–11, 12.1–34). And at the end of the
play, the king grants the shoemakers the boon of allowing them to have

two market days per week for buying and selling leather (21.151–62). The making of clothing, specifically shoes, however, is not only a source of money for the shoemakers: their craft is also a source of great pride, particularly for Eyre, who incessantly declares that shoemakers are gentlemen, and that he is nobly, even "princely" born (7.45–47, 10.147–48, 11.16–17, 21.17, 35)[8] and that his employees and fellow shoemakers are gentlemen (1.214–17, 7.45–46, 10.153–56, 20.1, 21.6–9, 146–47). Furthermore, the shoemakers do show solidarity in standing with Ralph to win back his wife from Hammon, justifying Hodge's description of his fellows as "the brave bloods of the shoemakers, heirs apparent to Saint Hugh, and perpetual benefactors to all good fellows" (18.1–3). Dekker confronts the upperclass belief that to work with one's hands is shameful when the Earl of Lincoln sneers at his nephew Lacy's stint as a shoemaker on the continent, "A goodly science for a gentleman / Of such descent!" (1.30–31);[9] however, Lacy himself decides that "The Gentle Craft is living for a man!" a statement later echoed by Eyre (3.24, 11.46). Since it is this craft and its practitioners that allow Lacy to win Rose, help Eyre become Lord Mayor of London, support Ralph in regaining Jane, and conduce to the final happy ending, it is hard not to see this trade as an unalloyed good to the commonwealth.

As the repeated passage of largely unenforced sumptuary laws makes clear, clothing was regarded as an important status marker during the late medieval and early modern periods,[10] and the equation of clothing and status is reflected in this play. In the first scene, Roger Oatley, although a knight and Lord Mayor of London, objects to his daughter marrying Lacy, nephew of an earl, because "Poor citizens must not with courtiers wed, / Who will in silks and gay apparel spend / More in one year than I am worth by far" (1.12–14). When Eyre dons the alderman's finery before meeting with the skipper whose cargo he wishes to buy, his foreman Hodge describes the probable reaction to such dress: "I warrant you, there's few in the city but will give you the wall, and come upon you with the 'Right Worshipful'" (7.114–16). Eyre's wife, Margery, is also very impressed with Eyre's transformation: "By my troth, I never liked thee so well in my life, sweetheart" (7.121–22). Eyre's appropriation of clothing above his station apparently helps him complete a business transaction that brings him so much wealth that he actually does rise in station. When word comes to his wife that he will be named sheriff of London, she immediately starts to think of ordering new clothes (10.30–37). When Eyre's journeyman Firk announces Eyre's appointment to her, he says, "on with your best apparel" (10.113–14). And when Eyre himself appears, he says to Margery, "I shall make thee a lady; here's a French hood for thee," adding, as they make ready to depart to dinner with the Lord Mayor, "Come, Madge, on with your trinkets" (10.139–40, 153).[11] Obviously, clothing is closely connected with improving and maintaining one's worldly status. Clothing can also represent a character's condition, as when Hodge tries to reassure Ralph by telling him that Jane has been seen in London "very brave and neat" (10.104). Firk even suggests that wearing "a gown and a gold ring" will make his master feel young again (7.148–50).

The value of clothing is also evident in its perceived use as an object of bribery. Eyre attempts to buy Ralph out of military service by offering the captains Lacy and his cousin Askew all the boots they will need for seven years (1.133–35). In an extravagant gesture, Rose Oatley offers her maid Sybil her cambric apron, Romish gloves, purple stockings, and a stomacher to go to London and learn whether Lacy is to be sent to France (2.52–56).

Clothing, however, is not always valued in this play. Sybil describes Roland Lacy dressed to go to war: "By my troth, I scant knew him—here 'a wore a scarf, and here a scarf, here a bunch of feathers, and here precious stones and jewels, and a pair of garters—O monstrous!—like one of our yellow silk curtains at home" (2.25–28).[12] She makes it clear that, in her opinion, this martial Lacy is excessively proud and unpleasant (2.30–42). Jane rejects the wealthy Hammon for the working-class Ralph by referring to Ralph in terms of his clothing:

> [To Ralph] Thou art my husband, and these humble weeds
> Makes thee more beautiful than all his wealth. (18.58–59)

Jane also attempts to represent her rejection of Hammon and his wealth through rejection of the clothing he has given her:

> Therefore I will but put off his attire,
> Returning it into the owner's hand,
> And after ever be thy constant wife. (18.60–62)

Clothing is also represented as simply trivial; Eyre urges Rose Oatley not to marry a courtier: "A courtier?—wash, go by! Stand not upon pishery-pashery. Those silken fellows are but painted images—outsides, outsides, Rose; their inner linings are torn" (11.39–42). Eyre also refers to the French hood he had earlier given Margery—and her (presumably new) farthingale—as "trash, trumpery, vanity!" (17.17). When Margery later tries to advise her husband how to address the king, Simon rebukes her in terms referring to clothing as insignificant: "Shall Sim Eyre learn to speak of you, Lady Madgy? Vanish, Mother Miniver-Cap, vanish! Go, trip and go, meddle with your partlets and your pishery-pashery, your flews and your whirligigs!" (20.51–55). On the other hand, clothing proves of ultimate importance to Ralph and Jane, when his recognition of the shoe he has made for her helps him locate and reclaim her as his wife, almost magically, just as she is about to marry another man.[13]

Several of the play's characters also use clothing references as analogies. When Lacy decides to disguise himself as a shoemaker, he announces his intention "to clothe his cunning with the Gentle Craft" (3.4). Eyre describes the Lord Mayor's life as "a velvet life" (17.38). The most frequent user of such analogies is Firk, who uses them to declare that he is leaving Eyre's employ: "'Nails, if I tarry now, I would my guts might be turned to shoe-thread'"; to describe Lacy and Rose's marriage: "they shall be knit like a pair of stockings in matrimony"; and to threaten Hammon's servants when they attempt to reclaim the wedding clothes that their master has given to Jane: "Bluecoat, be quiet. We'll give you a new livery else. We'll make Shrove Tuesday St. George's Day for you. . . . Touch not a rag, lest I and my brethren beat you to clouts" (7.57–58, 16.117–18; 18.70–71, 73–74).[14]

Finally, several of the proverbs quoted in the play involve clothing, invariably in a negative sense. Sybil pretends to address the absent Lacy, by saying, "thou mayst be much in my gaskins, but nothing in my netherstocks" (2.39–40), implying that she objects to his pride.[15] Sybil also, after accepting Rose's bribe of clothing to discover Lacy's plans, declares "I'll sweat in purple": her acceptance of expensive clothes will damn her to hell. When Eyre pretends to be an alderman, Firk suggests that his master "will be as proud as a dog in a doublet" (7.109–10).[16] Finally, in a kind of anti-clothing proverb, Margery, for all of her love of fine clothes, announces, "Naked we came out of our mother's womb, and naked we must return" (10.99–100).

The varying attitudes toward clothing represented in this play highlight some of the central conflicts of both the medieval and early modern periods. While the noblemen view shoemaking and other handwork as degrading, the workers themselves see it not only as necessary, but as—literally—potentially ennobling, since it is one of their own who becomes Lord Mayor of London in the course of the play. Clothing is also a signifier of wealth, which is, on the one hand, coveted, but on the other criticized as being antithetical to the values of Christianity. The workers in the clothing industry are in large part serving the upper classes, yet the play presents a strong sense of class conflict, with the workers ultimately triumphant.

In a play that is often interpreted as a charming romantic comedy—and which certainly is that, among other things—Dekker also examines questions of class, work, and the body, which clothing is designed to protect and adorn.[17] Within the bounds of comedy, Dekker grounds the play in serious and realistic concerns.

Notes

1. Patricia Thomson maintains that in *The Shoemaker's Holiday,* Dekker de-
   picts a relatively static feudal society rather than a more dynamic early
   modern one. Patricia Thomson, "The Old Way and the New Way in
   Dekker and Massinger," *Modern Language Review* LI (1956): 168–78, 169.
   L. C. Knights suggests that Dekker inherits his morality from the Middle
   Ages. L. C. Knights, *Drama and Society in the Age of Jonson* (New York:
   Charles W. Stewart, n.d.), 7–8. George R. Price maintains throughout his
   book that Dekker's sources, models, and views on social and economic
   problems are all medieval. George R. Price, *Thomas Dekker,* Twayne's English Authors Series 71 (New York: Twayne, 1969), 34–6, 51, 59, 68, 72–3,
   76–7, 79, 85, 128, 131, 136, 138–9, 157. Madeleine Doran suggests that
   Dekker's mixture of comedy and romance with more serious matter is an
   inheritance from medieval drama. Madeleine Doran, *Endeavors of Art: A
   Study of Form in Elizabethan Drama* (Madison: University of Wisconsin
   Press, 1954), 209–11.
2. Anthony Parr, ed., *The Shoemaker's Holiday* (The New Mermaids, 2d ed.
   1990; rpt. London and New York: A & C Black/W. W. Norton, 1994), xxi.
   All quotations from the play are taken from this edition. On "the feasting
   imagery," see also Harold E. Toliver, "*The Shoemaker's Holiday:* Theme and
   Image" (1961); rpt. in *Shakespeare's Contemporaries: Modern Studies in English Renaissance Drama,* 2d ed., ed. Max Bluestone and Norman Rabkin
   (Englewood Cliffs, New Jersey: Prentice-Hall, 1970, 184–93), 191–2.
3. See Eugene P. Wright, *Thomas Deloney,* Twayne's English Authors Series
   323 (Boston: G. K. Hall/Twayne, 1981), 40, 51, 56, 58–9.
4. Quoted in R. G. Howarth, *Two Elizabethan Writers of Fiction: Thomas Nashe
   and Thomas Deloney* (Cape Town: University of Cape Town, 1956), 41; see
   also Wright 72–3. See Wright 64–5 for the importance of the clothing
   trade spelled out in *Jack of Newbury,* and Wright 76–7, 79 for its importance
   in *Thomas of Reading.*
5. Wright 63, 60. For the importance of the clothing trade in England during this period, see Wright 52–3. Price notes that Simon Eyre was "historically a woolen-draper but made a shoemaker by Deloney" (51), and J. R.
   Sutherland enlarges on the biography of the real individual: "Eyre was an
   historical figure, an upholsterer and later a draper of London, who was
   Sheriff in 1434, Lord Mayor from 1445–6, and who died, a wealthy man,
   in 1459. In telling the story of Eyre's life, Deloney, who took various lib-
   erties with the facts, boldly changed him into a shoemaker. The confusion
   was natural, as Eyre the draper had built a Leadenhall in 1419, which, since
   the fifth year of Elizabeth's reign, had been used as a leather market."
   Dekker's Shoemaker's Holiday (Oxford:
   Clarendon Press, 1928), 8. Fredson Bowers, however, rejects the claim
   that Eyre built a Leadenhall. Fredson Bowers, ed. Cyrus Hoy, *Introductions,
   Notes, and Commentaries to texts in "The Dramatic Works of Thomas Dekker,"*

edited by Fredson Bowers. 4 vol. (Cambridge: Cambridge University Press, 1980), vol. 1: 15–16; and Paul C. Davies explains, "Leadenhall was in existence in the previous century. It was made over to the city in 1411 by the then Lord Mayor, Richard Whittington, and it was not until 1419 that Simon Eyre erected a public granary here, to which the old name was transferred." Paul C. Davies, ed., *The Shoemakers' Holiday* (Berkeley and Los Angeles: University of California Press, 1968), 102 n. 146.

6. In Deloney, Eyre can't afford to buy the ship's cargo, so he takes his wife's advice to "disguise himself as a rich Alderman and buy the cargo on credit" (Wright 90). On the morality of Eyre's wearing aldermanic attire and Dekker's purpose in including this ambiguous incident, see Julia Gasper: "If we wish, we can find in this ambiguity a well-observed truth about the business world. When he is dressed in his garded gown, Simon Eyre is told by Hodge, 'Why now you looke like your self master, I warrant you.' If Eyre poses as an alderman, he is taken for one, and so eventually becomes a real one. You are what you can persuade people you are. Credit is credit." Julia Gasper, *The Dragon and the Dove: The Plays of Thomas Dekker,* Oxford English Monographs series (Oxford: Clarendon Press, 1990), 32; see also 31. Gasper would presumably question M. C. Bradbrook's statement that Dekker "never shows cheating tradesmen." M. C. Bradbrook, *The Growth and Structure of Elizabethan Comedy* (London: Chatto & Windus, 1962), 122. For other examples of early modern characters using clothing to claim status that they don't really deserve, see Robert Greene's reference in his address "To the Curteous Reader" of *The Blacke Bookes Messenger* to the confidence man Ned Browne, who "was in outward shew a Gentlemanlike companion attyred very braue." Robert Greene, *The Blacke Bookes Messenger,* ed. G. B. Harrison. Elizabethan and Jacobean Quartos (1922–1926; rpt. New York: Barnes & Noble, 1966), 2. "Ned Browne" also tells of persuading one of his acquaintances to disguise himself as a constable, by means of "a faire cloake and a Damaske coate." Greene, *Messenger,* 9. Greene's "Roberto" in his *Groats-worth of Witte* remarks that a player's "outward habit" caused him to mistake the actor "for a Gentleman of great liuing." Robert Greene, *Greene's Groats-worth of Witte, bought with a million of Repentance* and *The Repentance of Robert Greene,* ed. G. B. Harrison. The Bodley Head Quartos (London: John Lane, 1923), 33. "Cuthbert Conny-Catcher" tells of wearing livery and a cognizance and "[bearing] the port of a Gentleman," and a marginal note states that "Some Conicatchers dare weare noblemens liueryes, as W. Bickerton and others." "Cuthbert Conny-Catcher," *The Defence of Conny-Catching,* ed. G. B. Harrison. Elizabethan and Jacobean Quartos (1922–1926; rpt. New York: Barnes & Noble, 1966), "To the Readers," 6; see also "Conny-Catcher," 33. "Cuthbert" also describes "a crew of terryble Hacksters in the habite of Gentlemen" (38) and a rogue who dresses his brother as a servant to allow himself to pose as a gentleman of means (42; see also 46, 48–50, and 58–60).

7. "[Lacy's] disguise as a shoemaker is in a sense his own creation of identity; paradoxically, only by becoming a shoemaker can he and Rose share in the "frolic, so gay, and so green, so green. . . . To this extent, his love takes on the qualities which modify the romantic spirit of the second song ["Trowl the bowl"]—his disguise is a symbolic acquisition of the sturdiness of the lower classes." Toliver, 187.

8. See also 17.20: "Prince am I none, yet bear a princely mind."

9. Gasper points out: "Certainly Dekker meant Lincoln's speech . . . on the shame of a gentleman practising a craft, to be deeply ironic. Lacey commands most respect when he is least a member of his 'lacie' family. He is ennobled by the gentle craft." Gasper, 28.

10. See Russ McDonald, *The Bedford Companion to Shakespeare: An Introduction with Documents* (Boston and New York: Bedford Books of St. Martin's Press, 1996), 234–5, 271.

11. As Thomas Marc Parrott and Robert Hamilton Ball point out, Margery "[rejoices] in the hood, the periwig, and the mask that mark her rise in social status; the 'world's calling is costly,' she says, 'but it is one of the wonderful works of God.'" Thomas Marc Parrott and Robert Hamilton Ball, *A Short View of Elizabethan Drama, Together with Some Account of Its Principal Playwrights and the Conditions under Which It Was Produced* (1943; rpt. New York: Charles Scribner's Sons, 1958), 109. Although their central point— Margery's delight in her new finery—is surely correct, their quotation is not, since what Margery really says is "Fie upon it, how costly this world's calling is! Perdie, but that it is one of the wonderful works of God, I would not deal with it" (10.49–51).

12. Bowers compares this description to "the vision of Rafe, done up as the Lord of May, in *The Knight of the Burning Pestle.*" Hoy, 33. Toliver sees Lacey's clothing as "characteriz[ing] his affectation," and maintains that Rose provides him with the opportunity to exchange this finery "for the true garland of festivity." Toliver, 187.

13. Peggy Fay Shirley points out that shoes are comically inappropriate as a lover's parting gift. Peggy Fay Shirley, *Serious and Tragic Elements in the Comedy of Thomas Dekker,* Salzburg Studies in English Literature: Jacobean Drama Studies 50 (Salzburg: Institut für Englische Sprache und Literatur, Universität Salzburg, 1975), 15–16, 27. On the other hand, "[Ralph's] gift of a pair of shoes, besides being practical, becomes a symbol of fidelity and humbleness. It offers a metaphorical language for the poor to talk about love without ostentation. . . . And by the shoes Ralph is enabled to find Jane as she is about to accept a countergift of Hammon's 'rings,' which will give her a chance to have 'lily hands' of grace rather than the working hands of craft. . . ." Toliver, 189. It is also Lacy's disguise as a shoemaker that allows him access to Oatley's house, where he and Rose plot the escape that leads to their marriage; Dekker appears to have been fond of the device of a man disguising himself to fit his lover with a pair of shoes, since he repeats it in *Match Me in London.*

14. Bowers comments: "Servants were traditionally permitted to change masters (and so livery) on St. George's Day. Blue was the usual color of servants' livery. The servant here, however, is promised a livery of black and blue and red, from the bloody beating Firke threatens to administer." Hoy, 64.

15. Bowers states, "Sybil is saying simply that, so far as she is concerned, Lacy, for all the fashionable elegance of his external appearance, is but an ordinary mortal." Hoy, 34.

16. That "a dog in a doublet" was, at least potentially, a ridiculous figure, is clear from William Harrison's use of the phrase in a passage discussing the English predeliction for mixed and changing fashions in clothes: "Such is our mutability that today there is none to the Spanish guise, tomorrow the French toys are most fine and delectable, ere long no such apparel as that which is after the High Almain [German] fashion, by and by the Turkish manner is generally best liked of, otherwise the Morisco [Moorish] gowns, the Barbarian sleeves, the mandilion worn to Collyweston-ward, and the short French breeches make such a comely vesture that, except it were a dog in a doublet, you shall not see any so disguised as are my countrymen of England." William Harrison, *The Description of England,* ed. George Edelen (1968; rpt. Washington, D. C. and New York: Folger Shakespeare Library/Dover, 1994), 145–6; referred to by Bowers, Hoy, 47.

17. Allardyce Nicoll describes it as "one of the most charming romantic plays of the period," adding, however, "yet we are forced to acknowledge that its charm rests rather on the surface than penetrates within." Allardyce Nicoll, *British Drama,* 5th ed. rev. (New York: Barnes & Noble, 1963), 94. But is not charm—like clothing—always a matter of surface?

CHAPTER 14

VALUE-ADDED STUFFS AND SHIFTS IN MEANING:
AN OVERVIEW AND CASE STUDY
OF MEDIEVAL TEXTILE PARADIGMS

Désirée Koslin

Throughout the medieval period in Europe, textile production and its adjunct industries was the lifeblood of medieval life and economics.[1] Textile trade and commerce brought revenue as they introduced novel technologies and design aesthetics for merchants to disseminate and for entrepreneurs to emulate.[2] The urban centers of Europe grew around the ports and places of textile manufacture, and many cloth merchants became rich and powerful in city governance and as suppliers to the courts. Surviving documents, especially trade accounts and inventories, allow an understanding of the direct relations between luxury consumption and the demands of ceremonial circumstance. For instance, massive expenditures to purchase precious textiles were incurred for royal or ecclesiastical investitures, and for the matrimonial and funerary requirements of medieval society's elite.[3] In prestigious commissions of works of art, the costliest pigments and elaborate techniques were used to depict these fine textile qualities and luxurious garments by artists who were employed by secular and ecclesiastical patrons of art.[4]

One aim of this chapter is to describe the many and varied aspects of medieval textile making, and to pay special attention to the finishing treatments that added value and status to fabrics through specialized labor, technology, and, at times, design novelty. Another goal is to address the significance, meaning, and problems of colors and types of medieval textiles in dress and in their idealized representations in written and pictorial sources. These issues will be considered both in a *longue durée* sense as well as through case studies. I will also propose that rhetorical issues are frequently encountered in representations of textiles and dress where they can be seen to have historicizing, polemical, and ideological intent.[5]

The labor-intensive nature of medieval textile production involved a great many people in society, with the largest number of contributors at its very lowest rungs. They were the agricultural laborers who tended the sheep, or grew and retted the flax while their women cleaned, carded, and combed the fleece, or hackled and scutched the flax before they spun these fibers into woolen, worsted, or linen thread.[6] Wool was by far the most important textile material, followed by linen and hemp. Together they made up the great majority of medieval textiles for clothing and domestic purposes—silk, seen in disproportionate numbers in representations and surviving medieval textiles, accounted for only a small fraction of the total. The weaving of woolen cloth and linen took place in domestic, rural/manorial settings during the earlier period, and predominantly in the urban workshops governed by the corporate guild structures during the later Middle Ages. Compensation paid to weavers was certainly a rung above that of spinners, but still did not amount to much.[7] Within the weavers' ranks, those making simple, narrow cloth appear to have worked at subsistence level, while others with specialized skills and advanced loom equipment were better off; these differences can be seen in medieval representations.[8] Toward the top of the labor pyramid of medieval textile manufacture are those employed in after-treatments that added value and quality, such as dyers, fullers, and shearers, while the merchants of cloth, dyes, and luxury textiles represented the pinnacle.[9] This simple picture is made more complicated, however, by the fact that these professional groups were not on equal social footing, or had clearly defined ranks. Weavers and fullers obeyed ordinances established by and imposed on them by the merchant guilds. Dyeing, often quite lucrative, was sometimes part of the merchants' own businesses—several cases of litigation reveal that some merchants even employed artisans to do both weaving and dyeing for them.[10]

Like the dyers of late antique Rome, who were organized into a *collegium tinctorum* of several ranks, medieval dyers were classified into categories by the types of dye stuffs they used. This also determined, for instance, where in the urban aquifer systems their establishments could be located, since dyeing was a smelly, polluting business.[11] In France and Italy, dyers continually fought with merchant weavers for independence, but it was only in the sixteenth century that this status was achieved and dyers finally had their own corporations independent of the merchants, allowing opportunities for technical specialization and scientific improvement of the craft. It is in this period that the first dye manuals appear in print, examples being *Tbouk va Wondre,* published in Brussels in 1513, and the popular *Plictho de l'arte de tentori,* published by Giovanni Ventura Rosetti in Venice, 1540/1548.[12]

Professional dyers had been active in the ancient cultures of China, India, and Egypt, and their methods were noted in early texts that sometimes conflate the dyers' occupation with that of alchemy and a notion of deceit.[13] This latter aspect has a moral dimension that reverberate in the prescriptions of the Church Fathers. Jerome, for instance, compares virginity to vice in a letter to young Laeta with a textile metaphor, "Once wool has been dyed purple, who can restore it to its previous whiteness?"[14] The later, reform-minded clerics also clamor for a return to simple, undyed fabrics for the virtuous clothing of the religious, rejecting the artifice and expense of dyed textiles. The public's apparent distrust of dyers may also refer to the practical difficulties in obtaining clear, strong, and lasting textile colors, and probably also to the reputed habit of dyers in adulterating or supplementing or replacing expensive dyes with lesser ones.[15]

Medieval sensibilities responded to brilliant, full-hued, and unambiguous color, epitomized by precious stones, metals, and fine pigments used for jewelry and precious objects, stained glass, and enamel.[16] These naturally colored substances had become imbued with absolute and symbolic meaning, derived from biblical references, such as Ezekiels's Vision of Eden, and from the Neoplatonic ideas about light current among intellectuals of the time.[17] These concepts can be said to take a manifest and more popular form during the twelfth century when heraldry was adopted and codified—no doubt a consequence of the encounter with the foe's martial signage during the recent Crusades. The heraldic *tinctures* comprised two metal colors, gold/yellow (*or*), and silver/white (*argent*), as well as the stains of red (*gules*), blue (*azure*), black (*sable*), green (*vert*), and purple (*purpure*).[18] When tinctures were applied to designate a positive purpose, they denoted auspicious symbolic values of long standing. *Or* stood for virtue and prestige; *argent* for purity and chastity; *gules* for humanity and the blood of martyrs; *azure* for eternal truth and the heavenly divine; *sable* for mourning and death; *vert* for hope, joy, and resurrection; and *purpure* for majesty.

The symbolic significance of these shiny, bright colors of metal and mineral content could certainly be transmitted into textile form, although the quality of the tinctures could not. The brilliant blue in the molten glass or on the artist's palette was prepared from azurite or the yet costlier, highest quality *lapis lazuli,* neither of which was soluble in water for use as a dye. Contracts for works of art often stipulate prices, quantities, and purposes for the best blue—the heavenly mantle of the Virgin Mary, for instance.[19] But, to reiterate the point, the intense *lapis* blue used for coloring the garments in the painted calendar pages of Duke John of Berry's *Très Riches Heures,*[20] could not have been attained in actual cloth using the available medieval dyestuffs. Blues from woad or indigo, although saturated and beautiful in their own right, did not match the *lapis* vividness and luminosity. Furthermore, natural dyes were limited by seasonal availability, strength, and consistency, and did not, for instance, include a direct, bright green—this color had to be achieved by top-dyeing yellow over blue. The original textile substrate also had to be available in as pure a state of whiteness as possible in order to take on bright tonalities from the dye. In terms of wool, this meant using select, fine fleeces without pigmentation. For plant fibers like flax and hemp a painstaking bleaching process must precede dyeing. To best emulate the idealized *tinctures,* expensive and rare, naturally white silk would be used for dyeing.

Symbolism of medieval color was not one-dimensional, however. The positive value was reversed when a tincture was applied to a negatively charged subject, for example, a prostitute's red headdress, a Jewish or Oriental woman's yellow gown, or the frivolous and sometimes vulgar display of parti-colored garments.[21] Such dialectical contrasts and mutual incompatibilities abound in medieval imagery and literature, dichotomies that today are keeping a number of scholars at work on unveiling new social and theoretical constructs. One must, of course, presume that the medieval reality presented far fewer complexities in the sartorial signals and in the actual, available textiles than those we encounter in the surviving works of art. Still, and because textiles were so valuable, a colored garment was perceived as a carrier of meaning. At a glance it conveyed a complex message to the informed viewer: a visual identification of the value and color quality of the fabric, its social significance and appropriateness, and its abstract, metaphorical color associations. In addition, the garments' length, width, number of layers, and quality of lining material completed such an instant evaluation.

Medieval color concepts also had spiritual dimensions, and the anagogical function of color in precious substances and objects was described as an important accessory in medieval devotional practices.[22] Here again, the reverse also applies, seen in the abstention from color in the dress of the medieval religious who followed prescribed ideals set out by the early founders,[23] following a long tradition of ascetic precepts of humility and rejection of worldly preoccupations. Furthermore, their garments of "white," "black," and "gray" conformed by no means to absolute color terms or values, but instead constituted convenient, well-understood designations for the different monastic and conventual orders. There are, for instance, several orders of White monks and nuns current from the eleventh century onward that include the Cistercian, Gilbertine, Camaldolese, Olivetan, and Humiliati/ae orders. They rejected the black, expensive cloth of their Benedictine "parents" or peers by reverting to the undyed, unprocessed woolens considered to be following more closely the prescriptions of their rule's founder, St. Benedict of Nursia (ca. 480–ca. 550). Their "whites," therefore, are frequently depicted in a range of light neutral shades, as from a random crop of natural fleeces, all the while conveying a symbolic impact of "white" as pure, chaste, and humble.[24] The Grey Friars, or Franciscan brethren and sisters, are depicted in a range of dark neutral colors including grays and browns, indicating that *bure,* the simplest, coarsest, and least processed of the available woolen fabrics, was used for their meanly cut tunics.

Like the textiles discussed so far, the majority of medieval fabrics were woolen and made of solid, natural, or dyed colors. Their qualities ranged from coarse and open light-weights that include the lesser grades of *bure, beige, biffe,* and *tiretaine,* and medium qualities like *camelin,* to the superb *brunettes* and scarlets, soft, dense, and velvety.[25] It is the latter that we see rendered in the convincing illusionism of the fifteenth century's new medium of oil on panel paintings in Northern Europe. Scarlet's elaborate finishing method, thrice-repeated brushing and shearing the raised surface of this superior woolen quality, had been a specialty in the Low Countries at least since the eleventh century, but we can of course only vicariously appreciate it through the verisimilitude of these later representations.[26] The name scarlet signals the red color derived from *kermes,* an insect dye collected in Asia Minor and the most expensive of medieval dyes. By the late medieval period, this prestigious color name applies to the exquisite woolen quality regardless of color, a shift in nomenclature not uncommon in the marketing of luxurious commodities.

Rare and costly silk textiles with pattern-woven decorative and figured designs had long been produced in the various cultural and political entities around the Mediterranean, or obtained through trade to the east before, during, and after the Crusades. The migration of Muslim weavers from east to west ultimately brought advanced textile technology to Italy, enabling its nascent silk industry to develop during the thirteenth century.[27] It has often been noted that textile designs and their compositional schemes followed long-standing traditions, suggesting conservatism on behalf of the weavers. However, these professionals were, just like their counterparts today, merely in the business to meet the demands of their clients, who clearly preferred the traditional designs. The authority of the roundel scheme with its enclosed imagery of stylized animals lasted for some six hundred years, conveyed to its patrons, perhaps, ideas of royal symbology stemming from ancient fables. (Figure 14.1.) By the thirteenth century, the roundel was replaced by a rich variety of new design schemes. With or without the roundel framework, textile designs that are symmetrically oriented on the vertical axis

Figure 14.1   Woven Textile, Eastern Mediterranean, eleventh or twelfth century. 20-³/₁₆ x 12-¹³/₁₆ in. (51.2 x 32.6 cm). New York, Cooper-Hewitt, National Design Museum, Smithsonian Institution/Art Resource, NY. Gift of John Pierpont Morgan. 1902–1–122.

have shown much staying power, responding unconsciously, perhaps, to the upright and paired symmetry of the human body itself.

These roundel designs in figured silks from the first millennium of the Common Era were woven either in the weft-faced compound plain weave, *taqueté,* or compound twill, *samit.*[28] Since the early centuries weaving technology to produce these sophisticated weave structures had been in development in China, Western Asia, and the Eastern Mediterranean. At first, presumably, simple pattern heddle rods were operated by the weaver, and by degrees pattern-selecting devices were introduced, fitting the loom with lifting mechanisms activated by a "draw boy" who assisted the weaver.[29] Designs of astounding size and complexity were created on such draw looms in the imperial workshops of Constantinople, epitomized by the eleventh-century Elephant Silk, found in the tomb of Charles the Great (d. 814) in Aachen. Its original width on the loom is calculated to have been ca. 240 cm, and the compound twill structure displays the color combination favored in Byzantium of a red-purple ground with majestic elephants in blue, white, yellow, and dark green, set in pearl roundels, each 78 cm in diameter.[30]

Polychrome silks, such as the Aachen Elephant example, were wondrous not only for their technological sophistication, but also for the superior skills demonstrated in the preparatory stages when the silks were yarn-dyed in the different colors prior to weaving. Compared with the piece dyeing process described above, in which entire fabrics lengths were immersed in dye vats and colored a solid shade, the dyeing of the individual hanks of silk with uniform results before the weaving process was much more exacting. The fastness of each color was of paramount importance; no excess dye could be permitted to remain since such dye migration would doubtless later compromise the fabric in which these colors were to be combined. Of course, it would be particularly challenging to feature very dark colors with very light ones in the same textile. The dyeing of black, for instance, required repeated over-dyeing of different colors to achieve depth, and it could later be prone to "crocking," migration of excess dye particles to the lighter areas of a yarn-dyed textile.

For the weaving of the Elephant Silk, two weavers would have been seated on opposite sides of the very wide loom, passing the five shuttles with different colors back and forth between them in unvarying synchronization and precision, thousands of times per roundel repeat. Meanwhile, the draw boy had to put in a flawless performance as well, maintaining his grip on the lashes, or strings of pattern selection, for each of the sequences of the five weft colors, to then pull forth a new set of lashes in a probably numbing repetitiousness. Designs on the scale of the Aachen Elephant Silk were extremely rare, and the great majority of roundel-type silks, so prevalent from the sixth to the twelfth centuries, feature much smaller dimensions.

The hegemony of the vertical symmetry in textile design was interrupted briefly by a new, asymmetrical style introduced during the upswing of east-west trade during the Yuan Dynasty (1279–1368) in China, and it influenced styles across the Asian mainland. Under the Mongol Ilkhans of Iran and Iraq, a fusion of Western, Middle Eastern, and East Asian aesthetics was encouraged in the Tabriz manuscript workshops and artisan ateliers under Rashid al-Din (d. 1318), vizier to Sultan Uljaytu (r. 1304–16).[31] The Italian textile designers exposed to these new currents adopted Chinese motifs, characterized by animated plumage, foliage, and landscape motifs, and inserted these freely into the prevailing Gothic styles of decoration.[32] The narrative content of these lengths of silk was not unlike that in the scenes of the margins in many books of hours of the thirteenth and fourteenth centuries, which were populated by hybrid animals, incongruous spatial relationships, pseudo-Kufic inscriptions, and abundant, obviously private jokes.[33] With their graphic emphasis, these "marginal" textiles generally had fewer and lighter colors than the previous roundel style. They usually featured a metallic, gilt-silver thread, and were often executed in the novel *lampas* technique that allowed more than one weave structure or texture to be seen on the face of the fabric.[34] The figured fabrics so prominently depicted in the costumes of thirteenth- and fourteenth-century paintings and in manuscript illumination can be presumed to illustrate this silk quality.[35]

This witty and light-hearted style was supplanted by stately and aggrandizing "pomegranate" floral designs during the fifteenth century, when the most luxurious ones are rendered in silk velvet, frequently having supplementary gilt-silver wefts worked in a variety of glittering surface treatments. The pomegranate motif had its roots in ancient Mesopotamian iconography as a symbol of life and fertility through its display of a fruit with many seeds.[36] Central Asian textile versions of the pomegranate had reached Europe during the Yuan period; the motif was also taken up by Ottoman weavers in Bursa and Istanbul on a grand scale during the fifteenth century. Splendid pomegranate textiles had place of pride in Italy's silk cities of Florence, Genoa, and Venice, as well as in Spain's Valencia and Seville. This constitutes another example of the pan-Mediterranean fusion referred to earlier in the roundel-framed designs popular through the twelfth century. The pomegranate and other giant, symmetrical floral motifs will continue to have a strong presence in status clothing, decorative panels, and cloths of estate/honor through the first half of the eighteenth century.[37]

Medieval artists depicted historical and biblical events in their own present, using contemporary settings and dressing the figures in garments of the day to transmit visual messages primarily through clothing cut, shape,

Figure 14.2   *Saint Andrew with Scenes from His Life* (retable), c. 1420–30. Tempera on wood, gold ground. H. 123 1-¼ W. 123- ⅝ in. (313.1 x 314 cm). New York, The Cloisters of The Metropolitan Museum of Art, 06.1211.1–2, Rogers Fund.

design, and color. Within this framework, however, and with knowledge of the successive styles of textile designs, the artist and/or his client also made efforts to establish historical perspectives by introducing past styles. A Catalan retable, ca. 1420–30, now in The Cloisters of The Metropolitan Museum of Art, is an important example of the regional painting style associated with Luis Borrassá.[38] (Figure 14.2) It depicts scenes from the life of St. Andrew, and displays textiles in many of its panels. Most of the textiles are in the contemporaneous fifteenth-century style, but a consciously archaizing example is also shown. In one of the *predella* scenes, St. Andrew is conducted by a woman to the bedside of her ailing sister, whose bedspread is rendered in an archaic style featuring large pearl roundels. This is a conscious effort to render the settings around the disciple of Christ with textiles evoking this distant past, whereas the cloth of honor behind the

Figure 14.3 *The Hunt of the Unicorn as an Allegory of the Passion: The Unicorn Leaps out of the Stream*. Tapestry, Southern Netherlands, 1495–1505. Wool and silk with silver and silver-gilt threads. From the Chateau of Verteuil. New York, The Metropolitan Museum of Art, The Cloisters Collection. 37.80.5. Gift of John D. Rockefeller, Jr.

Virgin Mary in the center panel displays the contemporary large floral motifs implying continuous, eternal time.

By the late Middle Ages, a wide range of textile qualities was available to clothe the members of the rapidly diversifying society. The color, shape, and fit of dress items might proclaim, betray, or feign social status beyond the three estates, and artists were encouraged to find ways to express these nuances. This was a task they set out to achieve with apparent relish and much ingenuity by stereotyping and caricaturing their subjects for an un-ambiguous and swift visual impact. An informed medieval viewer could also, like the student of reception theory today, search understated details for clues involving inscriptions, gesture, ornament details, facial features, and so on, for still more subtle messages.

In the series of four tapestries recently identified as *The Hunt of the Unicorn as an Allegory of the Passion*, also in The Cloisters collection, many of these markers are present.[39] In all but the last of the four, men only are depicted. They are busy in the act of hunting, the prerogative of medieval noblemen, and are armed with spears and swords; some carry hunting horns in simple or elaborately decorated shoulder holsters. Under attack, the unicorn kicks back and rears its virginal body against the men's thrusting spears in scenes set in the *mille fleurs* surround of seigneurial pleasures, resonating powerfully with the imagery of violence.

In the second tapestry in this suite (Figure 14.3), the highest-ranking member of the hunting party is easily picked out by the prominence of his plumed headdress; his garments, as can be expected, are made of expensive red silk velvet and pomegranate-patterned brocade.[40] The lord's brocade jacket, worn over a sleeved velvet doublet, is lined with another textile, as befits his class. By contrast, the "lymerers" in charge of the greyhounds wear plain-colored, mostly unlined, and presumably woolen clothing. The other men in the hunting party appear in a variety of sleeved or sleeveless short jackets over tight-fitting, solid color or striped red-and-white hose, and shirts as undergarments are glimpsed. An ostentatiously dressed young man, perhaps a younger son of a noble family in search of his fortune, displays dress features of a provocative nature in the neatly tied front closure of the hose, a precursor to the codpiece. Usually this detail is a sign of uncouth rusticity, encountered in late medieval genre depictions of peasants at labor, such as wine harvest, or in rambunctious dance.[41]

In the Cloisters' *Hunt,* several of the men's jackets, sleeves, or linings are rendered in bold, wood grain–like textile patterns in bright colors—this is watered silk, or *moiré.* Its shifting reflections of light were caused by a treatment given to the tightly woven, ribbed silk involving wetting and great pressure that resulted in a partially flattened weave texture, a method known to have been practiced already in Abbasid Baghdad (ca. 750–1258).[42] The second tapestry in the *Hunt* series shows the unicorn attacked by hunters with spears. They have exaggerated, coarse facial features, and clothing that is provocatively tight, rendered in garish moiré and in strongly contrasting colors that agree with the many characteristics seen in the Tormentors of Christ scenes so thoroughly examined by Ruth Mellinkoff.[43]

The *Hunt* series presents a comprehensive view of the custom of slashing the upper garment to expose the one underneath. Martial circumstances and practicality account for this display—a garment outgrown, or acquired too small, perhaps in booty, could be made to fit a mercenary's shoulders by strategically placed slashes like the ones seen in the third figure from the left. Long hanging sleeves were also slashed along the front so they could be thrown back, allowing the arms to move freely. The hunter about to plunge his spear into the unicorn's neck wears a blue, unlined moiré jacket so short and tight that its lower edge has ripped and curled, and the sleeve seams appear to have burst open—male desire turned into fashion convention.

A great vogue for moiré is apparent in late medieval tapestry representations from the Brussels workshops.[44] From ca. 1480 until 1525, moiré appears in the tapestry medium either in martial contexts, or worn by foreigners and those of doubtful reputation, or used for dressing allegorical figures that portray negative characteristics. For example, armed merchants or mercenaries wear moiré cowls and shirts in a tapestry depicting the Infancy and Upbringing of Romulus and Remus.[45] Bathsheba's husband Uriah, destined for the battlefield, wears a red moiré doublet, while King David's blue mantle is lined with green moiré.[46] Various allegorical figures in moiré are seen with the sword-yielding personification Wrath wearing armor, and a turbaned figure beneath her wears moiré as well in a tapestry probably made in Brussels.[47] A Magdalen-like figure with red hair appears in green moiré in a four-part tapestry series of the Life of St. John the Baptist.[48] A tapestry fragment depicting the Three Fates with female personifications of diverse illnesses includes Delirium wearing a moiré mantle.[49] The fabric type also appears in the famous Lady and the Unicorn series, in which the subordinate maidservant wears blue or red moiré gowns. When personifying Vision, the Lady herself wears blue moiré under her velvet-lined brocaded gown. She holds a mirror, symbol of the mortal sin of vanity, in which she captures the reflection of the unicorn who is in repose on her lap.[50]

Moiré, with its shifting and unstable textile aspects, is used in the woven works cited above to indicate negative values in the allegorical and biblical figures. The fabric is not seen depicted in painted works of art of the period, and may be due to issues allied to the art medium in question. Tapestry lends itself very easily to interpreting the highlights and random patterns of moiré through its discontinuous wefts and color hatching technique. The methods used by painters to render regularly repeating textile patterns included tracings, stencils, and stamps, but these may not have been suitable for producing the more arbitrary and shifting moiré meanders. It is noteworthy that in the pen and wash drawings by Bernard Van Orley (ca. 1488–1541), made for the Romulus and Remus tapestry cited above, no fabric textures or patterns are present.[51] One of the several anonymous masters or workshops active in Brussels at the time could well have turned the moiré texture into a specialty, and reused it time and again as a textile trope for a state of ambiguity that the fabric itself so prominently embodied. Others may then have adopted it for less specific purposes. Moiré therefore provides useful material for a case study to conclude this chapter linking the production of textiles in the Middle Ages with the rhetorical schemes and dichotomous messages that fabrics and clothing represent in medieval culture.

Notes

1. An excellent, recent synthesis is by Dominique Cardon, *La Draperie au Moyen Âge: Essor d'une grande industrie europénne* (Paris: CNRS Éditions, 1999). For the later Middle Ages, the essays in N.B. Harte, ed., *The New Draperies in the Low Countries and England, 1300–1800* (Oxford and New York: The Pasold Research Fund and Oxford University Press, 1997) are most useful.

2. For the exchanges with Islamic Spain, the Mediterranean region, and West and East Asia, see Karel Otavsky and Muhammad Abbas Muhammad Salim, *Mittelalterliche Textilien: Ägypten, Persien und Mesopotamien, Spanien und Nordafrika* (Bern: Abegg-Stiftung Riggisberg, 1995); Riggisberger Berichte Band 5, *Islamische Textilkunst des Mittelalters: Aktuelle Probleme* (Bern: Abegg-Stiftung Riggisberg, 1997); and James C.Y. Watt and Anne E. Wardwell, *When Silk Was Gold: Central Asian and Chinese Textiles* (New York: The Metropolitan Museum of Art, 1997).

3. See, for instance, Agnes Page, *Vêtir le Prince. Tissus et couleurs à la cour de Savoie (1427–1447),* (Cahiers Lausannois d'histoire médiévale 8, Lausanne, 1993); Lisa Monnas, "Textiles for the Coronation of Edward III" in *Textile History,* Vol. 32, No.1 (2001); 2–35, and Françoise Piponnier, "Les Étoffes de deuil" in *À réveiller les morts* (Lyon: Presses Universitaires, 1993), 135–40.

4. For this context, see Jonathan J. G. Alexander, *Medieval Illuminators and Their Methods of Work* (New Haven and London: Yale University Press, 1992), especially 52–71; also Odile Blanc, "Parures Sacrées" in *Brocarts Celestes* (Avignon: Musées du Petit Palais, 1997), 23–30; and Michael Baxandall, *Painting and Experience in Fifteenth-Century Italy* (Oxford and New York: Oxford University Press,1972).

5. See, for instance, J. J. G. Alexander, "*Labeur et Paresse*: Ideological Images of Medieval Peasant Labour" in *Art Bulletin* 72 (1990): 436–52.

6. Cardon, *La Draperie au Moyen Âge: Essor d'une grande industrie europénne,* meticulously details these preparatory stages, 145–301.

7. Ibid.; for a brief summing up of the few facts available on the topic, see 600–6.

8. A well-known example of the former is the ca. 1250 colored pen drawing of a weaver depicted naked at his loom, weaving simple *saye* (Cambridge, Trinity College Library, Ms. O.9.34, f.32v). By contrast, the many stained-glass windows contributed to churches by weavers' guilds depict respectably dressed and shod men, working in pairs at broadcloth looms, in France for instance at Notre-Dame, Semur-en-Auxois; Amiens Cathedral; St. Etienne, Elbeuf; and in Belgium in the basilica of St. Martin, Halle.

9. See John H. Munro, "The Medieval Scarlet and the Economy of Sartorial Splendour," in *Cloth and Clothing in medieval Europe,* eds. N. B. Harte and K. B. Ponting (London: The Pasold Fund, 1983), 13–70.

10. See E. M. Carus-Wilson, *Medieval Merchant Ventures: Collected Studies* (London: Methuen, 1954), 226–33.

11. See Étienne de Boileau, *Livre de métiers,* ca. 1260, listing dyers of *bon teints* producing fast dyes, and *petit teints,* less durable ones. Later documents from Germany and Italy also maintain this division.

12. See *Förädlad Textil* (Processed Textiles), Borås, Sweden, 1951; and the 1968 facsimile of G.V. Rosetti, *Plichto de Larte de Tentori,* Venice 1540/1548.

13. Herodotos, Dioscurides, and Pliny the Elder describe dye recipes with scientific, empirical intent. Plutarch and the anonymous, third-century author of the *Papyrus graecus holmiensis,* Uppsala, University Library, include references to alchemy, see R. J. Forbes, *Studies in Ancient Technology,* Vol. IV (Leiden: Brill, 1956), 128–36.

14. Joan M. Petersen, ed. and trans., *Handmaids of the Lord: Contemporary Descriptions of Feminine Asceticism in the First Six Centuries* (Kalamazoo: Cistercian Publications, 1996), 258.

15. See Michel Pastoreau, "Jésus Tenturier. Histoire symbolique et sociale d'un métier réprouvé," in Médiévales No. 29 *L'Étoffe et le vêtement* (1995), 47–64. In this fascinating essay, Pastoreau's claim that top-dyeing was not practiced in the Middle Ages is untenable in view of the many records and surviving objects that attest to its frequency.

16. See Michael Camille, *Gothic Art: Glorious Visions* (New York: Harry N, Abrams, 1996), esp. 41–57; and John Gage, *Color and Meaning: Art, Science and Symbolism* (Berkeley and Los Angeles: University of California Press, 1999), esp. 68–81.

17. Exemplified by Abbot Suger of St. Denis (ca. 1081–1151), and Bishop Robert Grosseteste of Lincoln (ca. 1168–1253).

18. See A. R. Wagner, *Heralds and Heraldry in the Middle Ages* (London: HMSO, 1967); and Ottfried Neubecker, *Wappenkunde* (Munich: Orbis, 1991); and Michel Pastoreau, *Couleurs, images, symboles. Études d'histoire et d'anthropologie* (Paris: 1989).

19. See Daniel V. Thompson, *The Materials and Techniques of Medieval Painting* (New York: Dover, 1956), 134.

20. Chantilly, Musée Condé, Ms. 1695.

21. See Ruth Mellinkoff, *Outcasts: Signs of Otherness in Northern European Art of the Late Middle Ages,* 2 vols. (Berkeley, Los Angeles, and Oxford: University of California Press, 1993) for a full treatment of the late medieval material, and among others by Michel Pastoreau, "Formes et couleurs du desordre: le jaune avec le vert," in *Médiévales* 4 (1983): 62–73. Then as now, brightly contrasting colors would also be used for visual attention-getting, as in medieval heraldry and livery and contemporary traffic signs.

22. See the oft-cited passage in Abbot Suger's "De Administratione" in Erwin Panofsky, ed., *Abbot Suger on the Abbey Church of St.-Denis and Its Art Treasures* (Princeton: Princeton University Press, 1946), 57–67.

23. See my dissertation, *The Dress of Monastic and Religious Women as Seen in Art from the Early Middle Ages to the Reformation* (New York: New York

University Press, Institute of Fine Arts, January 1999) for a comprehensive treatment.

24. In medieval secular society, the wearing of chaste all white was by special, clerical dispensation only, as we learn in Margery of Kempe's account of her life; see W. Butler-Bowdon, ed., *The Book of Margery of Kempe* (London and Toronto: Oxford University Press, 1936), 97, 108, 134, and elsewhere.

25. For terms see Françoise Piponnier and Perrine Mane, *Se vêtir au Moyen Âge* (Paris: Adam Biro, 1995); Elisabeth Hardouin-Fugier et al., *Les Étoffes: Dictionnaire Historique* (Paris: Les Éditions de l'Amateur, 1994), and A. R. Bridbury, *Medieval English Clothmaking: An Economic Survey* (London: The Pasold Fund, 1982).

26. See John. H. Munro, "The Medieval Scarlet and the Economy of Sartorial Splendour," 13–70.

27. For an exceptionally lucid and concise account, see Priscilla P. Soucek, "Artistic Exchange in the Mediterranean Context," in *The Meeting of Two Worlds: The Crusades and the Mediterranean Context,* ed. Clifton Olds (Ann Arbor: The University of Michigan Museum of Art, 1981), 15–16, and the textile entries of this catalogue.

28. For a concise introduction, see Agnes Geijer, *A History of Textile Art: A Selective Account* (London: Pasold Research Fund, 1979), 57–60.

29. See Luther Hooper, *Hand-Loom weaving: Plain and Ornamental* (London: Pitman & Sons, 1910 (1953)); John Becker, *Pattern and Loom: A Practical Study of the Development of Weaving Techniques in China, Western Asia and Europe* (Copenhagen: Rhodos, 1987); and Regula Schorta, "Zur Entwicklung der Lampastechnik," in *Riggisberger Berichte Band 5* (1997): 173–80.

30. See Anna Muthesius, *Byzantine Silk Weaving AD 400 to AD 1200* (Vienna: Verlag Fassbaender, 1997), 38–9, and catalogue M58.

31. See Sheila S. Blair and Jonathan M. Bloom, *The Art and Architecture of Islam 1250–1800* (New Haven and London: Yale University Press, 1994), 25–33.

32. See Monique King and Donald King, *European Textiles in the Keir Collection: 400 BC to 1800 AD* (London and Boston: Faber and Faber, 1990), 44–54; and Anne E. Wardwell, "*Panni Tartarici:* Eastern Islamic Silks Woven with Gold and Silver (13th and 14th Centuries)" in *Islamic Art* III (1989): 95–174.

33. See Alexander (1992), 118. Toward the end of the seventeenth century, another exoticizing and eclectic textile style, bynamed "bizarre," will evoke the spirit of these "marginal" examples.

34. On this late nineteenth-century term, see Geijer, *A History of Textile Art: A Selective Account,* 60.

35. See Brigitte Tietzel, *Italienische Seidengewebe des 13., 14. und 15. Jahrhunderts* (Cologne: Deutsches Textimuseum Krefeld, 1984).

36. See Friedrich Muthmann, *Der Granatapfel: Symbol des Lebens in der Alten Welt* (Bern: Schriften der Abegg-Stiftung, 1982).

37. Versions of the pomegranate pattern have continued, in a historicizing mode, right up to the present as a "period" furnishing textile style, showcased by most decorating firms in their residential lines.

38. Rogers Fund, 1906, 06.1211.1–9.

39. The Cloisters' two other tapestries and two fragments depicting unicorn motifs are now thought to belong to other pictorial cycles; see Adolfo S. Cavallo, *Medieval Tapestries in The Metropolitan Museum of Art* (New York: The Metropolitan Museum of Art, 1993), 297–327, catalogue numbers 20 b, c, d, e; and his *The Unicorn Tapestries at The Metropolitan Museum of Art* (New York: The Metropolitan Museum of Art, 1998).

40. Technically, "brocade "designates extra-weft patterning, and can be applied to any woven substrate. Here it is used in its popular sense, designating a richly patterned, compound weave, usually including metallic threads, but distinct from velvet, for instance, in that it doesn't feature any pile. See subject listed in Dorothy Burnham, *Warp and Weft: A Textile Terminology* (Toronto: Royal Ontario Museum, 1980).

41. See several tapestry examples in Fabienne Joubert, *La tapisserie médiévale au musée de Cluny* (Paris: Réunion des Musées Nationaux, 1994), 93–103. On the voyeuristic depictions of the lower classes, see the previously cited article by J. J. G. Alexander, n. 5, and Ruth Mellinkoff's *Outcasts*, n. 21, as well as Keith Moxey's various works, for instance his "Sebald Beham's Church Anniversary Holidays: Festive Peasants as Instruments of Repressive Humour," in *Simiolus 12* (1981–82): 107–30.

42. Later in the Modern Era, this treatment is reported in 1640 being applied to ribbed worsted qualities as well, see Eric Kerridge, *Textile Manufactures in Early Modern England* (Manchester: Manchester University Press, 1985), 53.

43. See Ruth Mellinkoff, *Outcasts: Signs of Otherness in Northern European Art of the Late Middle Ages,* especially chapter 6.

44. This fabric's close relative, changeable or shot silk, is less easy to identify in depictions; see discussion in Cage, *Color and Meaning: Art, Science and Symbolism,* 51–2.

45. Guy Delmarcel, *Golden Weavings: Flemish Tapestries of the Spanish Crown* (Malines, Munich and Amsterdam: Gaspard De Wit Foundation, 1993), cat. no. 10, c. 1525–1530, The Foundation of Rome, now La Granja de San Ildefonso of the Collection of the Spanish Crown.

46. Delmarcel, *Golden Weavings: Flemish Tapestries of the Spanish Crown,* cat. no. 3. David receives Bathsheba in his Palace, ca. 1515, Brussels workshop, now Madrid Palacio Real.

47. See Alan Phipps Darr et al., *Woven Splendour: Five Centuries of European Tapestry in the Detroit Institute of Arts* (Seattle and London: The Detroit Institute of Arts, 1996), ca. 1500–1510, now in the Detroit Institute of Arts collection, cat. No. 9.

48. See Cecilia Paredes et al., *Âge d'or bruxellois: Tapisseries de la Couronne d'Espagne* (Brussels: Bruxelles/Brussel, 2000), cat. No. 10–13, ca. 1515–1520, Brussels workshop, now Madrid, Palacio Real.

49. See Joubert, *La tapisserie médiévale au musée de Cluny,* cat. no. XII a) The Triumph of Honor, and b) The Fates, now in Cluny Museum, Paris, 140–9.

50. Joubert, *La tapisserie médiévale au musée de Cluny,* cat. no.VI, 66–92.
51. See Delmarcel, *Golden Weavings: Flemish Tapestries of the Spanish Crown,* 66–7. Clearly, the professional who carried out the enlargement to full-scale cartoon size from the artist's conception must have contrived a way to indicate the pattern features to the weavers.

GLOSSARY

Ar—Arabic; E—English; F—French; G—Greek; Ir—Irish; It—Italian; L—Latin; OF—Old French; Pe—Persian.

beige   F; coarse, loosely woven woolen cloth of lesser quality.

biffe   F; coarse, loosely woven woolen cloth of lesser quality.

biretta   It; from L *biretus,* a soft cap, in wool or silk with origins in the *pileus.* In the Middle Ages, worn by some professional ranks, e.g., doctors, judges, and so on.

bliaut, bliaut gironé   OF.; orig. term for costly silk, by mid-twelfth century designates courtly outer (silk) garment for both men and women. A two-piece variant for women displays lacing at sides and multiple horizontal folds, the skirt, *gironé,* richly pleated and trailing.

brat   Ir; lower body garment, rectangular cloth fastened with a pin.

brocade   E; generic term for patterned luxury textile, usually with metallic content. Structurally, a weave with added, supplementary weft elements (threads) for decorative purposes.

buckram   E; during medieval period, a term for fine, delicate linen or cotton quality, by the late period it designates a coarse linen, stiffened with glue or starch.

bure, burel   F; coarse woolen cloth of lesser quality.

cambric   E; sixteenth-century term for fine white linen, originally made in Cambrai.

ceinture   F; (silk) girdle, often decorated with applied metal plaques or studs, embroidery or patterned weaves.

chemise   F; from L *camisia;* see tunic.

cloth of estate   E; a richly figured textile panel suspended behind a secular, important individual. Historical persons are frequently depicted, and royal inventories take up such cloths.

cloth of honor   E; in art, a richly figured textile panel suspended behind a divine personage. The pattern repeats are displayed as prodigiously large. Weave structures are presumed to be *lampas* or *samitum,* not velvet, which was used for clothing only in this period.

cornet, -te   F; development and extension of the female hood's tip and flaps, some reaching the ground.

cors, corps   OF; upper part of gown, usually with slits and lacing to achieve a tight fit.

cowl   E; hood, of varying dimensions. In fourteenth century, the hood's tip develops into a long tippet, and the collar is rolled to form a brim, this is the fashionable *liripipe.*

cuirass(e)   F; rigid, two-piece armor joined to protect chest and back.

cotte   F; E; *kirtle,* tunic- based garment worn over shift, and under *surcot.* Lacing and buttoning allowed form-fitting styles, amply represented in the art of the fourteenth century

coudière   F; fourteenth-century long sleeves hanging empty from elbows in both male and female versions, a fashionable and frivolous dress detail.

damask   E; from thirteenth century, a simple (one set of warps, one set of wefts) weave structure of turned satin textures (warp emphasis countered by weft emphasis), capable of elaborate figured design as well as simple stripes and checks. Silk was the original quality, later worsted and linen were used.

dorser, dosser   E; a plain or figured fabric covering the back of a seat, throne, or seating area.

draw boy   E; weaver's assistant, see draw loom.

draw loom   E; a sophisticated weave technology with patterning capacity to render elaborate, repeated designs. With origins in the East and Middle East, draw looms were used in Byzantium and Islamic Spain before being brought to Europe.

doublet   E; from early fourteenth century, a short jacket for men, seen in many forms: sleeved, sleeveless, with or without short skirts. See *pourpoint.*

fermail   F; brooch, clasp, fastener for cloaks, mantles, and so on.

ganache, garnache   F; I; fourteenth-century outer, wide, sleeveless, then cap-sleeved garment lined with fur or contrasting color lining, belted, and with matching headgear. An adaptation of thirteenth-century Muslim courtly dress style.

garter   E; band, sometimes finely decorated, tied just under knee to keep hose in place, worn by men and women. Elasticity in g. could be achieved by using braided (oblique interlacing) weave structure.

gaskins   E; late sixteenth-century hose or breeches

glass smoother   E; smooth piece of glass or stone used to flatten, gloss, and/or pleat textiles, especially damp linen fabrics.

girdle   E; sash or belt used by men and women, religious and secular. Visible girdles were often richly decorated with embroidery, metal plaques, etc. The religious must wear a girdle at all times over the coarse, woolen undergarment, prescribed by St. Benedict.

gonelles   F; from I, *gonna,* a term of eleventh-century origin denoting an outer garment worn by all, and based on the tunic.

gore   E; triangular panel set in straight-cut garments for more ample, flaring proportions. Gores reach the lower hem of the garment, whereas a *gusset,* seen for instance under the arm, does not.

half basket weave   E; an extended plain weave in which weft passes over and under groups of two warps, alternating each weft insertion. Full basket weave has paired wefts and warps in extended plain weave.

hauberk   OF, ME; defensive armor first around neck and shoulders, by twelfth century a full-length ring or chain mail.

hemp   E; cellulosic fiber, coarser than linen, used for simpler bed linen, work clothing, and so on.

hose   E; leg coverings, sometimes with feet, later joined at uppers to form, in effect, breeches. Medieval hose have been found, sewn of bias-cut woolen cloth.

houpplande   F; from mid-fourteenth-century first a male, then also a female outer garment of ample, rich folds. At first full length, open front with wide sleeves, with standing collar and side slits, by the fifteenth century shorter and fitted. Women's version develops a fur-trimmed "v" neckline, and the wide, hanging sleeves display a great variety of exaggerated shapes.

greenweed   E; *genista tinctoria,* a plant yielding yellow and green dye color.

jaque, jaquette F; from late thirteenth century, the garment worn over armor, the term has acquired varied and shifting meaning up to the present. In fourteenth century it is longer than the *pourpoint,* reaching the thigh.

jerkin E; sixteenth-century term for men's jacket, longer than the doublet, often worn over it.

kermes from AR, PE, *qirmis;* dye substance, in "grains" obtained from insect *Coccus ilicis* and yielding a rich red, hence *"carmine," "cramoisy," "crimson."*

khil'a A; male robe of honor, caftan-style garment bestowed in ceremony by an Islamic ruler on his subjects.

lampas F; nineteenth-century term for compound weave structure introduced during twelfth century and eclipsing *taqueté* and *samitum* as the favored medium for figured silk weaves. *Diasper* is used in medieval sources for this structure that requires two sets of warp, and two or more sets of weft. Both warps are visible, and may display different weave structures (e.g., satin and twill), a feature not possible to achieve in earlier compound weaves with their "hidden" warps.

léine Ir.; sleeveless upper-body garment, sometimes hooded.

linen E; cellulosic fiber produced especially in Northern Europe for domestic and clothing uses, especially for table and bed linen, undergarments and veils. Medieval industries developed in the Low Countries specializing in weaving and bleaching linen.

madder E; *Rubia tinctorum,* a plant whose roots yield red dye color.

manicottoli It; see *coudière.*

mi-parti F; use of contrasting dress colors, termed *"demi parti"* in fourteenth century. Men's hose, or men's and women's gored tunics (*cottes*) in *mi-parti* signal perhaps merely innocuous or foppish high fashion, but frequently they appear as markers for those marginalized in medieval society, e.g., foreigners (especially Muslims and Jews), entertainers, prostitutes, and so on.

moiré F; watered silk; a treatment involving great pressure on folded, wetted, ribbed silk resulting in shifting reflections of light.

nålebinding N; looped structure that precedes knitting, executed with a large needle carrying a woolen thread.

nap E; the raised and shorn fibers of a fabric, brushed repeatedly with handle set with teasels. This process makes the textile softer, warmer, and denser.

netherstocks   E; late sixteenth-century stocking.

pileus   L; head gear with rolled brim, antique origins, made of felt, leather, wool, straw, or more costly materials.

pinbeater   E; small or large wooden or bone stick made smooth and used for pressing weft into place firmly while working the warp-weighted loom, and ancient weave technology.

plain weave   E; basic term for the most elemental of weave structures, in unvarying, alternating over-under sequence. Synonyms include linen weave, tabby, cloth weave. Fabric types of many weights designate plain weave, i.e., poplin, taffeta, gros-grain, rib, ottoman, and so on.

poulaines   F; shoe with extremely attenuated, pointed toes, fashionable from mid-fourteenth century through fifteenth century.

pourpoint   F; originally a jacket worn under armor, then transitioning from the mid-fourteenth century into a courtly, high-fashion style, requiring corseting and extreme postures. Hose were attached by laces (points) to the pourpoint. Originally padded and quilted for use under armor. *Paltock, jupon* are terms used for the generic jackets of the time.

robe   F; E; an ensemble of four to six (male) matched garments consisting of tunic, *cotte, surcot,* mantle, as well as hose and head gear.

S spun   E: carded and/or combed fibers, twisted in a "clockwise" direction—the resulting thread displaying an S-like slant. Linen is sometimes spun using S-direction of twist.

samit, samitum   derived from G *hexamitum,* a compound twill weave structure with two warp systems (one "hidden") allowing two or more wefts to create figured designs. It is the predominant structure for luxury textiles through the twelfth century.

scarlet   E; a) costly red dye color obtained from *kermes* insect; b) luxurious woolen cloth of any color produced in the Low Countries from the early medieval period. Napping and shearing may be repeated several times to achieve a buttery soft cloth.

shears   E; early cutting tool with two blades joined, good for cutting straight pieces (or shearing sheep), while the shank scissors (known from ninth century, but common from fourteenth century) with riveted blades allow curving cutting lines.

siglaton   F; a variant form of term designating a precious cloth, obsolete by 1400, defined variously as "scarlet," "cloth of gold," a fine figured fabric.

silk   E; luxury protein fiber from East Asia, limited production (sericulture) in
early medieval Europe, e.g., in Islamic Spain and Byzantium. From thirteenth
century, silk weaving centers established in Italy, notably in Lucca, Florence,
Venice, and Genoa.

slashing   E; the cutting of straight or curved incisions at edges, sometimes in the
body of clothing. Garments worn underneath are revealed. Martial origins have
been proposed, as in too-small garments obtained as booty that were slashed to
fit the new wearer.

sleeve   E; "bagpipe sleeve," "poky sleeve," "pudding sleeve" are descriptive terms
for the huge *houpplande* sleeves, with fitted cuffs.

surcot, -te   F; for men and women from twelfth century, a garment worn over the
tunic-based, long-sleeved *cotte*. By thirteenth century, deep armscyes develop,
and become enlarged so that by the fourteenth-fifteenth centuries women's
hips, waist, and upper arms are visible in the tight-fitting *cotte*.

tabard   F; full-length, usually male outer garment of varying forms, with or without sleeves, in its simplest form a rectangular piece of cloth worn similarly to a
poncho.

tablet weaving   E; ancient weaving method utilizing square tablets with warp
threads held in holes in each corner of the tablet. The result is a structure in
which the warp threads twine, and are held in place by wefts. It was portrayed
as fine ladies' work, and costly tablet-woven girdles survive.

tapestry   E; textile wall hanging. Earlier usage makes no distinction as to technique (embroidery, appliqué, weave), while today tapestry is defined structurally
as a woven textile with discontinuous wefts. Medieval tapestry is frequently narrative, on a monumental scale.

taqueté   F; term for compound plain weave structure with two warp systems (one
"hidden") allowing patterning in two or more colors. For comparison, see
samit.

tester   E; a plain or figured fabric suspended as a canopy above a bed, seat, or
throne.

tiraz   A; a) textile workshops supported by Islamic rulers, producing fabrics for
court and ceremonies; b) textiles with inscriptions produced in tiraz workshops,
especially honorific decorative bands.

tiretaine   F; coarse woolen cloth of mediocre quality, sometimes mixed with
linen.

tunic   E; from L *tunica,* a straight-cut, sleeved T-shaped garment worn by men and women from the late antique period onward. The term is used in medieval texts to designate the ubiquitous basic garment made of hemp or linen for underwear, and in wool for main, outer clothing.

under-dress, smock   E; shift, especially woman's, of linen or hemp.

velvet   E; cf F; "*velours,*" a costly and technologically advanced weave structure in which a supplementary warp forms loops protruding from the ground weave. The loops may be cut or left uncut, and the pile could be of different heights. The finest medieval velvets included supplementary, metallic wefts creating loops (*bouclé*) or flat, scintillating textures.

warp-weighted loom   E; one of the earliest weave technologies, utilizing warp threads suspended from a frame, kept taut with clay weights. Weaving proceeded from the top downwards. This loom is depicted in archaic Greek vases, and survived into the twentieth century in remote areas in Northern Europe.

weld   E; *Reseda Luteola,* a plant yielding a yellow dye color.

wool   E; protein fiber from sheep, carded, spun, and woven over most of medieval Europe. Major production centers in Low Countries, Lombardy, and Spain. The English wool sack was exported to the continent for spinning and weaving.

wool comb   E; handle set with long metal spikes, used in pairs to align choice woolen fibers for production onto worsted thread.

worsted   E; choice, long staple sheep's wool carded and combed for superlative quality thread.

Z spun   E; carded and/or combed fibers, twisted in a "counter-clockwise" direction—the resulting thread at times clearly displays a Z-like slant. Wool and cotton are usually spun in Z-direction.

CONTRIBUTORS

LINDA ANDERSON is Associate Professor, Department of English, Virginia Polytechnic Institute and State University. She is the author of *A Kind of Wild Justice: Revenge in Shakespeare's Comedies* (University of Delaware Press, 1987) and coeditor (with Janis Lull) of *A Certain Text: Close Readings and Textual Studies on Shakespeare and Others in Honor of Thomas Clayton* (University of Delaware Press, 2002). Anderson is presently working on a book about servants and service in Shakespeare's plays, and another about generational conflict in Renaissance drama.

ODILE BLANC is Researcher at Institut d'Histoire du livre of the École Nationale des Sciences de l'Information et des Bibliothèques (ENSSIB). Her publications include "Histoire du costume: quelques réflexions méthodologiques" (*Histoire de l'art, septembre* 2001), *Parades et parures. L'invention du corps de mode à la fin du Moyen Age* (Gallimard, 1997), "Le pourpoint de Charles de Blois: une relique de la fin du Moyen Age" (*Bulletin du CIETA* n.74, 1997), *Brocarts célestes,* exhibition catalogue, Musée du Petit-Palais, Avignon (1997), "Images du monde et portraits d'habits. Les recueils de costumes à la Renaissance" (*Bulletin du bibliophile,* 1995/2).

DONNA M. COTTRELL is an independent scholar in Shaker Heights, Ohio. She received her Ph.D. in Art in 1998 at Case Western Reserve University, and her dissertation is entitled: "Birds, Beasts, & Blossoms: Form & Meaning in Jan van Eyck's Cloths of Honor." Her paper, "Jan van Eyck's Closet Iconography," was presented at the 33rd International Congress on Medieval Studies, Kalamazoo, MI, in May 1998. Her research investigates Jan van Eyck's use and representation of textiles, including the cloth of honor in his Antwerp *Madonna by the Fountain.*

NINA CRUMMY is an archeologist small finds specialist. Her publications include *The Roman small finds from excavations in Colchester 1971–9,* Colchester Archaeological Report 2, 1983 (repr. 1995); *The coins from excavations in Colchester 1971–9,* Colchester Archaeological Report 4 (1987); *The post-Roman small finds from excavations in Colchester 1971–85,* Colchester Archaeological Report 5 (1988); *Excavations of Roman and later cemeteries, churches and monastic sites in Colchester, 1971–88,* Colchester Archaeological Report 9, with P. Crummy, and C. Crossan (1993); *Small finds from the suburbs and city defences,* with P. Ottaway and H. Rees, (Winchester City Museums publication 6, forthcoming). She is currently researching the deposition of typologies and chronologies of artifacts of Late Iron Age, Late Roman, and Late

Saxon periods, and medieval and post-medieval small finds from sites including Hertford and Stanway, Colchester.

BONNIE EFFROS is Associate Professor, Department of History, at State University of New York at Binghamton. She was the Sylvan C. Coleman and Pamela Coleman Memorial Fund Fellow (2001–2002) at The Metropolitan Museum of Art. She is the author of *Caring for Body and Soul: Burial and the Afterlife in the Merovingian World* (Penn State University Press, in press), *Merovingian Mortuary Archaeology and the Making of Early Medieval Europe* (University of California Press, in press); and *Creating Community with Food and Drink in Merovingian Gaul* (Palgrave Macmillan, forthcoming 2002). Effros is currently working on a project addressing the rise of interest in early medieval antiquities in France, Germany, and the United States in the late nineteenth and early twentieth century.

GLORIA THOMAS GILMORE is Associate Professor of French and Spanish at Westminster College, Salt Lake City, Utah. Her publications include "*Le Roman de Silence:* How Heldris Warps the Weave of Women's Work, of Textiles Telling Tales," in *Arthuriana* (forthcoming); "*Le Roman de Silence:* Allegory in Ruin or Womb of Irony," in *Arthuriana,* Summer 1997; "The Non-Sense of Plato's Body Bias: Existence as Metaphoric Praxis," in *Utah Foreign Language Review* (1995); "Marie de France Considers Conflicting Codes of Conduct: Equity in *Equitan,*" *UFLR* (June 1993); and "Foucault's Object, the Subject," in *UFLR* (May 1992). In her current research, Gilmore analyzes how textiles relate violence and subjectivity in the body of the *Lais* of Marie de France.

ELIZABETH WINCOTT HECKETT is Research Associate and part-time Lecturer in the Department of Archaeology, University College Cork, Ireland. Her publications include *Viking Age Headcoverings from Dublin, Medieval Dublin Excavations 1962–81* (National Museum of Ireland/Royal Irish Academy, 2002); "Archaeological Textiles from 13th–14th century Irish towns," in *Ypres and the Medieval Cloth Industry in Flanders,* eds. M. Dewilde, A. Ervynck and A. Wielemans (Instituut voor het Archeologisch Patrimonium, 1998); "Textiles, Cordage, Basketry and Raw Fibre" in *Late Viking Age and Medieval Waterford Excavations 1986–1992,* eds. M. Hurley and O. Scully with S. McCutcheon (Waterford Corporation, 1997); "Textiles" in C. Walsh, *Archaeological Excavations at Patrick, Nicholas and Winetavern Streets, Dublin* (Brandon Book Publishers Ltd., 1997); with R. Janaway, "The Textiles, Animal Hair and Yarn," in *Excavations by D.C. Twohig At Skiddy's Castle and Christ Church, Cork 1974–77,* eds. R. Cleary, M. Hurley and E. Shee Twohig (Department of Archaeology, UCC and Cork Corporation, 1997). Heckett's continuing research on Irish medieval textiles and dress includes further work on Irish archaic dress, and on a newly found hoard of Viking Age silver dress ornaments and silk textile.

SARAH-GRACE HELLER is Assistant Professor of French at Ohio State University. Her articles appear in *Speculum* and *Medievalia et Humanistica.* She is currently working on a book on the growth of the medieval French fashion system and the role that literature played in that system during the thirteenth century. She is also engaged in studies on sumptuary laws, Old French Crusade literature, and gender and embroidery.

PENNY HOWELL JOLLY is Professor of Art History and William R. Kenan Chair of Liberal Arts at Skidmore College in Saratoga Springs, New York. She has published articles in the *Art Bulletin, Burlington Magazine,* and elsewhere on Flemish and Italian artists, including Jan van Eyck, Rogier van der Weyden, Jacques Daret, and Antonello da Messina. She is the author of *Made in God's Image? Eve and Adam in the Genesis Mosaics at San Marco, Venice* (University of California Press, 1997). Her current work focuses on gender issues surrounding pregnancy, birthing, and breast-feeding, as depicted in Northern Renaissance painting.

DÉSIRÉE KOSLIN is Assistant Professor at Fashion Institute of Technology, where she teaches in the graduate Museum Studies: Costume and Textiles Department. Her publications include, "Under the Influence: Copying the *Revelaciones* of St. Birgitta of Sweden," in *Festschrift for Jonathan Alexander,* eds. Erik Inglis, Gerald Guest, Susan L'Engle (forthcoming 2002), "The Robe of Simplicity: Initiation, Robing, and Veiling of Nuns in the Middle Ages," in *Robes and Honor, The Medieval World of Investiture,* ed. Stuart Gordon (New York: Palgrave, 2001); "Manifest Insignificance: The Consecrated Veil of Medieval Religious Women," in *Sacred and Ceremonial Textiles, Proceedings of the Fifth Biennial Symposium of the Textile Society of America 1996;* "Structural and Stylistic Features in the Ritual Textiles of the Treasures of Dubrovnik," in *Treasures of Dubrovnik,* ed. Gabriel M. Goldstein, Yeshiva University Art Museum, New York, 1999; "Norse and Scandinavian Aspects in the Bayeux Embroidery," in *Ars Textrina* vol. 19 (1993); and "Turning Time in the Bayeux Embroidery," *Textiles and Text* vol. 13 (1990).

SUSAN L'ENGLE is Assistant Curator, Manuscripts Department, J. Paul Getty Museum, Los Angeles, California. Her publications include essays and catalogue entries in *Illuminating the Law: Medieval Legal Manuscripts in Cambridge Collections,* (Harvey Miller/Brepols, 2001), "your worthy servant Filitiana," in *Siena e il suo territorio nel Rinascimento,* vol. III, ed. Mario Ascheri (Edizioni il Leccio, 2000); "Trends in Bolognese Legal Illustration: The Early Trecento," in Acts of the Conference: *Juristische Buchproduktion im Mittelalter,* 25–28 October 1998, special issue of the periodical *Ius Commune* (Max-Planck-Institut für Europäische Rechtsgeschichte, forthcoming 2002); and "Outside the Canon: Graphic and Pictorial Digressions by Artists and Scribes," in *Festschrift for Jonathan Alexander,* ed. Erik Inglis, Gerald Guest, Susan L'Engle (forthcoming 2002). L'Engle's research efforts are devoted to medieval legal manuscripts and their iconography; she is currently working on an international survey of illuminated manuscripts of Roman law.

JANET SNYDER is Assistant Professor in the Division of Art at West Virginia University. Her publications include "The Regal Significance of the Dalmatic: the robes of *le sacre* represented in sculpture of northern mid-twelfth-century France," in *Robes and Honor, The Medieval World of Investiture,* ed. Stuart Gordon (Palgrave 2001); "Costumes in the Portfolio of Villard de Honnecourt," in *Villard's Legacy: Studies in medieval technology, science and art in memory of Jean Gimpel,* ed. Marie-Thérèse Zenner (Ashgate, 2002); "Knights and Ladies at the Door: Fictive Clothing in Mid-Twelfth-Century Sculpture," in *AVISTA Forum Journal* (Winter, 1996); "*'Bring me a soldier's garb and a good horse'*: Embedded stage directions in the dramas of Hrotsvit of Gandersheim," *Anthology of the Symposium, Hrotsvit of Gandersheim,* eds. Phyllis Brown and Linda McMillin (University of Toronto Press, in press). Snyder participates in the Limestone Sculpture Provenance Project. With Robert Bridges and Kristina Olson she is coediting *BLANCHE LAZZELL: An American Modernist* (West Virginia University Press, forthcoming 2003).

MAGGIE MCENCHROE WILLIAMS is an independent scholar. Her publications include "Warrior Kings and Savvy Abbots: The Sacred, the Secular, and the Depiction of Contemporary Costume on the Cross of the Scriptures, Clonmacnois," *Avista Forum Journal* 12/1 (Fall 1999); "Constructing the Market Cross at Tuam: The Role of Cultural Patriotism in the Study of Irish High Crosses," in *From Ireland Coming: Irish Art from the Early Christian to the Late Gothic Period and Its European Context,* ed. Colum Hourihane (Princeton University Press, 2001). Williams's current project, a revision of her *The Sign of the Cross: Irish High Crosses as Cultural Emblems,* considers the notion of copying and commodifying medieval objects in the pursuit of national identity, both in modern Ireland and in the Irish diaspora.

MARGARITA YANSON has completed her Ph.D. exams in Comparative Literature and Medieval Studies at University of California at Berkeley. She holds Bachelor's degrees from the University of Santa Cruz, and from St. Petersburg, Russia. The main focus of her dissertation will be the role of saintly characters in the literature of the High Middle Ages.

INDEX

Printed in the United States
129634LV00001B/3/P